Soft Computing Methods for System Dependability

Mohamed Arezki Mellal
M'Hamed Bougara University, Algeria

A volume in the Advances in
Systems Analysis, Software
Engineering, and High Performance
Computing (ASASEHPC) Book Series

Published in the United States of America by
    IGI Global
    Engineering Science Reference (an imprint of IGI Global)
    701 E. Chocolate Avenue
    Hershey PA, USA 17033
    Tel: 717-533-8845
    Fax: 717-533-8661
    E-mail: cust@igi-global.com
    Web site: http://www.igi-global.com

Library of Congress Cataloging-in-Publication Data

Names: Mellal, Mohamed Arezki, editor.
Title: Soft computing methods for system dependability / Mohamed Arezki
   Mellal, editor.
Description: Hershey, PA : Engineering Science Reference, 2020. | Includes
   bibliographical references. | Summary: ""This book explores the
   development and applications of soft computing methods in solving system
   dependability problems"--Provided by publisher"-- Provided by publisher.

Identifiers: LCCN 2019032915 (print) | LCCN 2019032916 (ebook) | ISBN
   9781799817185 (h/c) | ISBN 9781799817192 (s/c) | ISBN 9781799817208
   (eISBN)
Subjects: LCSH: Reliability (Engineering) | Soft computing.
Classification: LCC TA169 .S655 2020 (print) | LCC TA169 (ebook) | DDC
   620/.00452028563--dc23
LC record available at https://lccn.loc.gov/2019032915
LC ebook record available at https://lccn.loc.gov/2019032916

This book is published in the IGI Global book series Advances in Systems Analysis, Software Engineering, and High Performance Computing (ASASEHPC) (ISSN: 2327-3453; eISSN: 2327-3461)

British Cataloguing in Publication Data
A Cataloguing in Publication record for this book is available from the British Library.

All work contributed to this book is new, previously-unpublished material.
The views expressed in this book are those of the authors, but not necessarily of the publisher.

For electronic access to this publication, please contact: eresources@igi-global.com.

Advances in Systems Analysis, Software Engineering, and High Performance Computing (ASASEHPC) Book Series

ISSN:2327-3453
EISSN:2327-3461

Editor-in-Chief: Vijayan Sugumaran, Oakland University, USA

MISSION

The theory and practice of computing applications and distributed systems has emerged as one of the key areas of research driving innovations in business, engineering, and science. The fields of software engineering, systems analysis, and high performance computing offer a wide range of applications and solutions in solving computational problems for any modern organization.

The **Advances in Systems Analysis, Software Engineering, and High Performance Computing (ASASEHPC) Book Series** brings together research in the areas of distributed computing, systems and software engineering, high performance computing, and service science. This collection of publications is useful for academics, researchers, and practitioners seeking the latest practices and knowledge in this field.

COVERAGE

- Metadata and Semantic Web
- Computer Graphics
- Distributed Cloud Computing
- Engineering Environments
- Virtual Data Systems
- Computer Networking
- Performance Modelling
- Software Engineering
- Storage Systems
- Enterprise Information Systems

IGI Global is currently accepting manuscripts for publication within this series. To submit a proposal for a volume in this series, please contact our Acquisition Editors at Acquisitions@igi-global.com or visit: http://www.igi-global.com/publish/.

Titles in this Series

For a list of additional titles in this series, please visit:
https://www.igi-global.com/book-series/advances-systems-analysis-software-engineering/73689

Formal and Adaptive Methods for Automation of Parallel Programs Construction Emerging Research and Opportunities
Anatoliy Doroshenko (National Academy of Sciences of Ukraine, Ukraine) and Olena Yatsenko (National Academy of Sciences of Ukraine, Ukraine)
Engineering Science Reference • copyright 2020 • 195pp • H/C (ISBN: 9781522593843) • US $195.00 (our price)

Crowdsourcing and Probabilistic Decision-Making in Software Engineering Emerging Research and Opportunities
Varun Gupta (University of Beira Interior, Covilha, Portugal)
Engineering Science Reference • copyright 2020 • 182pp • H/C (ISBN: 9781522596592) • US $200.00 (our price)

Metrics and Models for Evaluating the Quality and Effectiveness of ERP Software
Geoffrey Muchiri Muketha (Murang'a University of Technology, Kenya) and Elyjoy Muthoni Micheni (Technical University of Kenya, Kenya)
Engineering Science Reference • copyright 2020 • 391pp • H/C (ISBN: 9781522576785) • US $225.00 (our price)

User-Centered Software Development for the Blind and Visually Impaired Emerging Research and Opportunities
Teresita de Jesús Álvarez Robles (Universidad Veracruzana, Mexico) Francisco Javier Álvarez Rodríguez (Universidad Autónoma de Aguascalientes, Mexico) and Edgard Benítez-Guerrero (Universidad Veracruzana, Mexico)
Engineering Science Reference • copyright 2020 • 173pp • H/C (ISBN: 9781522585398) • US $195.00 (our price)

Architectures and Frameworks for Developing and Applying Blockchain Technology
Nansi Shi (Logic International Consultants, Singapore)

For an entire list of titles in this series, please visit:
https://www.igi-global.com/book-series/advances-systems-analysis-software-engineering/73689

701 East Chocolate Avenue, Hershey, PA 17033, USA
Tel: 717-533-8845 x100 • Fax: 717-533-8661
E-Mail: cust@igi-global.com • www.igi-global.com

To my parents and my wife.

Table of Contents

Detailed Table of Contents

Vibha Verma, University of Delhi, India
Neha Neha, University of Delhi, India
Anu G. Aggarwal, University of Delhi, India

This chapter presents the application of grey wolf optimizer in software release planning considering warranty based on the proposed mathematical model that measures reliability growth of software systems. Hence, optimal release and warranty time is determined while minimizing the overall software development cost. The software cost model is based on failure phenomenon modelled by incorporating fault removal efficiency, fault reduction factor, and error generation. The model has been validated on the fault dataset of ERP systems. Sensitivity analysis has been carried out to study the discrete changes in the cost parameter due to changes in optimal solution. The work significantly contributes to the literature by fulfilling gaps of reliability growth models, release problems considering warranty, and efficient ways for solving optimization problems. Further, the grey wolf optimizer result has been compared with genetic algorithm and particle swarm optimization techniques.

Nabanita Banerjee, Techno Main Salt Lake, Kolkata, India
Sumitra Mukhopadhyay, Institute of Radio Physics and Electronics,
University of Calcutta, India

Noninvasive process of vital sign identification and design of low-cost decision-making system for the betterment of rural health care support is a prime facet of research. The identification of bio-signals from different sensors, noise removal, signal processing, and decision making requires the use of sophisticated expert system. In this chapter, the authors propose a modified multi grey wolf pack optimization technique (MMGWO) for better generalization and diversification. The basic model has been modified using net energy gain of the individual wolf in the process of hunting. Here multiple packs of wolves are considered with simultaneous sharing and conflict among them. The performance of the proposed technique is tested on 23 well known classical benchmark functions, CEC 2014 benchmark problem set along with classical real-life applications. The experimental results and related analysis show that the proposed MMGWO is significantly superior to other existing techniques.

Houda Daoud, ENIS, University of Sfax, Tunisia
Dalila Laouej, ENIS, University of Sfax, Tunisia
Jihene Mallek, ENIS, University of Sfax, Tunisia
Mourad Loulou, ENIS, University of Sfax, Tunisia

This chapter presents a novel telescopic operational transconductance amplifier (OTA) using the bulk-driven MOS technique. This circuit is optimized for ultra-low power applications such as biomedical devices. The proposed the bulk-driven fully differential telescopic OTA with very low threshold voltages is designed under ±0.9V supply voltage. Thanks to the particle swarm optimization (PSO) algorithm, the circuit achieves high performances. The OTA simulation results present a DC gain of 63.6dB, a GBW of 2.8MHz, a phase margin (PM) of 55.8degrees and an input referred noise of $265.3nV/\sqrt{Hz}$ for a low bias current of 52nA.

Hakduran Koc, University of Houston, Clear Lake, USA
Oommen Mathews, University of Houston, Clear Lake, USA

The unprecedented scaling of embedded devices and its undesirable consequences leading to stochastic fault occurrences make reliability a critical design and optimization metric. In this chapter, in order to improve reliability of multi-core embedded systems, a task recomputation-based approach is presented. Given a task graph representation of the application, the proposed technique targets at the tasks whose failures cause more significant effect on overall system reliability. The results of the tasks with larger fault propagation scope are recomputed during the idle times of the available processors without incurring any performance or power overhead. The technique incorporates the fault propagation scope of each task and its degree of criticality into the scheduling algorithm and maximizes the usage of the processing elements. The experimental evaluation demonstrates the viability of the proposed approach and generates more efficient results under different latency constraints.

Agriculture plays a vital role in the development of the Indian economy, and in addition, it contributes around 15% to the nation's GDP. Manually- or mechanically-operated diverse devices and supplies implied for farming machines are utilized in farming process. Still, sustainability is the most important issue in farming. Modern equipment smoke, dust, chemicals, and fertilizers both in manual-driven farming and modern farming are major environmental issues. So, in this chapter, sustainability issues in farming are studied, and a linear relationship between them can be found by interpretive structural modelling, such that the Micmac analysis and model can be developed for barriers of agricultural sector sustainability.

The main objective of this chapter is to analyze safety in railway systems through studying and understanding the train drivers' tasks and their common errors. Different approaches to classifying and analyzing driver errors are reviewed, as well the factors that affect driver performance. A comprehensive overview of the systems theoretic process analysis (STPA) method is presented, along with how it could be applied for controllers and humans. Quantitative risk assessment, along with some methods for quantifying human errors, are overviewed, and a Bayesian network is selected to study the effects of the identified driver errors. A case study aims to present a detailed quantitative safety analysis at European Train Control System (ETCS) system Levels 1 and Level 2, including driver errors. The STPA and Bayesian methods are combined to identify the hazards and quantify the probabilities of hazards when trains fail to stop at red signals.

Chapter 7

A distributed system is a complex system. Developing complex systems is a demanding task when attempting to achieve functional and non-functional properties such as synchronization, communication, fault tolerance. These properties impose immense complexities on the design, development, and implementation of the system that incur massive effort and cost. Therefore, it is vital to ensure that the system must satisfy the functional and non-functional properties. Once a distributed system is developed, it is very difficult and demanding to conduct any modification in its architecture. As a result, the quantitative analysis of a complex distributed system at the early stage of the development process is always an essential and intricate endeavor. To meet the above challenge, this chapter introduces an extensive framework for performability evaluation of a distributed system. The goal of the performability modeling framework is to consider the behavioral change of the system components due to failures. This reveals how such behavioral changes affect the system performance.

Chapter 8
Seongwoo Woo, Ababa Science and Technology University, Ethiopia

This chapter proposes how to decide the sample size. Due to cost and time, it is unfeasible to carry out the reliability testing of product with many samples. More studies on testing methods should explain how to decrease the sample size. To provide the accurate analysis for product reliability in the bathtub curve, Weibull distribution is chosen. Based on Weibayes method, we can derive the sample size equation from the product BX life. As the reliability of product is targeted and the limited sample size is given, we can determine the mission cycles from new sample size equation. Thus, by applying the accelerated loading to mechanical product, we can reveal the design failures of product and confirm if the reliability target is achieved. Comparing conventional sample size equation, this derivation of sample size equation is helpful to evaluate the lifetime of mechanical product and correct its design problems through parametric accelerated life testing (ALT). This equation also might be applicable to develop the test method of all sorts of structure in the civil engineering.

Chapter 9
Suchismita Satapathy, KIIT University, India

Occupational safety is a big issue of discussion for agricultural workers. The methods of working in the field in extreme climate totally depends on the environmental factor. Due to change in weather conditions, prices at the time of harvest could drop, hired labour may not be available at peak times, machinery and equipment could break down when most needed, animals might die, and government policy can change overnight. All of these changes are examples of the risks that farmers face in managing their farm as a business. All of these risks affect their farm profitability.

Heavy rains and drought could also damage or even wipe out crops. Another source of production risk is equipment. The most common sources of risk factor are weather, climate, diseases, natural disasters, and market and environmental factor shocks. Agricultural workers need sufficient precaution and safety measures at the time of field and machine work to minimize risk factor. So, in this chapter, an effort is taken to prioritize safety majors by MAUT method.

Foreword

This book provides a reference to graduate/postgraduate students, researchers, academicians and industrials. It includes a wide range of methods and applications that highlight soft computing methods.

Like the other books edited or written by Dr. Mohamed Arezki Mellal, this book is edited by a great attention.

Rekia Belhaoua
University of Claude Bernard 1, France

Preface

Technology in today's world has continued to develop into multifaceted structures. The performance of computers, specifically, has significantly increased leading to various and complex problems regarding the dependability of these systems. Recently, solutions for these issues have been based on soft computing methods; however, there lacks a considerable amount of research on the applications of these techniques within system dependability.

Soft Computing Methods for System Dependability is a collection of innovative research on the applications of these processing techniques for solving problems within the dependability of computer system performance. This book will feature comparative experiences shared by researchers regarding the development of these technological solutions. While highlighting topics including evolutionary computing, chaos theory, and artificial neural networks, this book is ideally designed for researchers, data scientists, computing engineers, industrialists, students, and academicians in the field of computer science.

The book is organized into nine chapters, written by experts and scholars from various countries. A brief description of each chapter is given as follows:

Chapter 1: This chapter presents the application of Grey Wolf Optimizer in Software Release Planning considering warranty based on the proposed mathematical model that measures reliability growth of software system. Hence optimal release and warranty time are determined while minimizing the overall software development cost. The software cost model is based on failure phenomenon modelled by incorporating Fault Removal Efficiency, Fault Reduction Factor and error generation. The model has been validated on fault dataset of ERP system. Sensitivity analysis has been carried out to study the discrete changes in the cost parameter due to changes in optimal solution. The work significantly contributes in the literature by fulfilling gaps of reliability growth models, release problems considering warranty and efficient ways for solving optimization problem. Further the Grey Wolf Optimizer result has been compared with Genetic Algorithm and Particle Swarm Optimization techniques.

Chapter 2: Noninvasive process of vital sign identification and design of low-cost decision-making system for the betterment of rural health care support is a prime facet of research now a days. The identification of bio-signals from different sensors, noise removal, signal processing and decision making requires the use of sophisticated expert system.

Method: In this chapter, we propose a modified multi grey wolf pack optimization technique (MMGWO) for better generalization and diversification. The basic model has been modified using net energy gain of the individual wolf in the process of hunting. Here multiple packs of wolves are considered with simultaneous sharing and conflict among them. The performance of the proposed technique is tested on 23 well known classical benchmark functions, CEC 2014 benchmark problem set along with classical real-life applications. The experimental results and related analysis show that the proposed MMGWO is significantly superior to other existing techniques.

Chapter 3: This chapter presents a novel Telescopic Operational Transconductance Amplifier (OTA) using the Bulk-Driven MOS technique. This circuit is optimized for ultra-low power applications such as biomedical devices. The proposed the Bulk-Driven fully differential Telescopic OTA with very low threshold voltages is designed under ±0.9V supply voltage. Thanks to the Particle Swarm Optimization (PSO) algorithm, the circuit achieves high performances. The OTA simulation results present a DC gain of 63.6dB, a GBW of 2.8MHz, a phase margin (PM) of 55.8 degrees and an input referred noise of 265.3nV/$\sqrt{\text{Hz}}$ for a low bias current of 52nA.

Chapter 4: The unprecedented scaling of embedded devices and its undesirable consequences leading to stochastic fault occurrences make reliability a critical design and optimization metric. In this chapter, in order to improve reliability of multi-core embedded systems, a task recomputation based approach is presented. Given a task graph representation of the application, the proposed technique targets at the tasks whose failures cause more significant effect on overall system reliability. The results of the tasks with larger fault propagation scope are recomputed during the idle times of the available processors without incurring any performance or power overhead. The technique incorporates the fault propagation scope of each task and its degree of criticality into the scheduling algorithm and maximizes the usage of the processing elements. The experimental evaluation demonstrates the viability of the proposed approach and generates more efficient results under different latency constraints.

Chapter 5: Agriculture plays a vital role in the development of Indian economy, and in addition to it contributes around 15% to the nation's GDP. Manually or mechanically operated diverse devices and supplies implied for farming machines are utilized in farming process. Still sustainability is the most important issue in farming. Modern equipment smoke, dust, chemicals and fertilizers both in manual driven farming and modern farming are major environmental issue.so in this paper sustainability issues in farming are studied and linear relationship between them can be found by interpretive structural modelling, such that the Micmac analysis and model can be developed for barriers of Agricultural sector Sustainability.

Chapter 6: The main objective of this chapter is to analyze safety in railway systems through studying and understanding the train drivers' tasks and their common errors. Different approaches to classifying and analyzing driver errors are reviewed, as well the factors that affect driver performance. A comprehensive overview of the systems theoretic process analysis (STPA) method is presented, along with how it could be applied for controllers and humans. Quantitative risk assessment, along with some methods for quantifying human errors, are overviewed, and a Bayesian network is selected to study the effects of the identified driver errors. A case study aims to present a detailed quantitative safety analysis at European Train Control System (ETCS) system Levels 1 and Level 2, including driver errors. The STPA and Bayesian methods are combined to identify the hazards and quantify the probabilities of hazards when trains fail to stop at red signals.

Chapter 7: A distributed system is a complex system. Developing complex systems is a demanding task when attempting to achieve functional and non-functional properties such as synchronization, communication, fault tolerance. These properties impose immense complexities on the design, development, and implementation of the system that incur massive effort and cost. Therefore, it is vital to ensure that the system must satisfy the functional and non-functional properties. Once a distributed system is developed, it is very difficult and demanding to conduct any modification in its architecture. As a result, the quantitative analysis of a complex distributed system at the early stage of the development process is always an essential and intricate endeavor. To meet the above challenge, this chapter introduces an extensive framework for performability evaluation of a distributed system. The goal of the performability modeling framework is to consider the behavioral change of the system components due to failures. This later reveals how such behavioral changes affect the system performance.

Chapter 8: This chapter proposes how to decide the sample size in the reliability design of mechanical system – automobile, refrigerator, etc. Due to the cost and time, it is unfeasible to carry out the reliability testing of product with many samples. More studies on testing methods therefore should be explained how to decrease the sample size in securing the product reliability. To reasonably provide the accurate analysis for product reliability in the bathtub curve, Weibull distribution is chosen. Based on Weibayes method defined as Weibull Analysis with a given shape parameter, we can derive the sample size equation from the product BX life. As the reliability of product is targeted and the limited sample size is given, we can determine the mission cycles from new sample size equation. Thus, by applying the accelerated loading to mechanical product, we can reveal the design failures of product and confirm if the reliability target is achieved. Comparing a variety of conventional sample size equation, this derivation of sample size equation is helpful to evaluate the lifetime of new mechanical product and correct its design problems through the parametric accelerated life testing (ALT). This equation also might be applicable to develop and select the test method of all sorts of structure in the civil engineering.

Chapter 9: Occupational safety is a big issue of discussion for Agricultural workers. The methods of working in field in extreme climate totally depends on environmental factor. Due to change in Weather conditions change; prices at the time of harvest could drop; hired labour may not be available at peak times; machinery and equipment could break down when most needed; draught animals might die; and government policy can change overnight. All of these changes are examples of the risks that farmers face in managing their farm as a business. All of these risks affect their farm profitability. Heavy rains and Draught without rain could also damage or even wipe out crops. Another source of production risk is equipment. The most common sources of risk factor are weather, climate, diseases, natural disasters, and market and environmental factor shocks. Agricultural workers need sufficient precaution and safety measures at the time of field and machine work, to minimize risk factor. So, in this chapter an effort is taken to prioritize safety majors by MAUT method.

Chapter 1
Software Release Planning Using Grey Wolf Optimizer

Vibha Verma
University of Delhi, India

Neha Neha
University of Delhi, India

Anu G. Aggarwal
ⓘD https://orcid.org/0000-0001-5448-9540
University of Delhi, India

ABSTRACT

This chapter presents the application of grey wolf optimizer in software release planning considering warranty based on the proposed mathematical model that measures reliability growth of software systems. Hence, optimal release and warranty time is determined while minimizing the overall software development cost. The software cost model is based on failure phenomenon modelled by incorporating fault removal efficiency, fault reduction factor, and error generation. The model has been validated on the fault dataset of ERP systems. Sensitivity analysis has been carried out to study the discrete changes in the cost parameter due to changes in optimal solution. The work significantly contributes to the literature by fulfilling gaps of reliability growth models, release problems considering warranty, and efficient ways for solving optimization problems. Further, the grey wolf optimizer result has been compared with genetic algorithm and particle swarm optimization techniques.

DOI: 10.4018/978-1-7998-1718-5.ch001

INTRODUCTION

In recent times IT-based firms focus on developing reliable software systems without bearing any financial or goodwill losses during the post-implementation phase. The technology advancements in Medical, Defence, Space, Transportation, banks, universities, homes appliances, etc. have increased the demand for qualitative software products. These products facilitate day to day task handling by reducing the efforts and time required at both the individual and organizational level and any failure encountered during the software operations may lead to heavy financial losses and sometimes may prove hazardous to human lives also (Yamada and Tamura, 2016).

Due to the ever-increasing importance of software systems, the researchers and IT firms are continuously working to improve their reliability and hence the quality. For this, it is very necessary to assess the reliability of a software system during its development phase. The software development process, also known as the software development life cycle, through this phase developers try to enhance the software quality. Here the development process comprises of few steps i.e., planning, analysis, design, implementation, testing and maintenance. Among all these steps, testing is considered as the most decisive and essential task for improving the quality of the software by detecting and removing the faults. Faults detectability influences software reliability growth which results in the development of Software Reliability Growth Models (SRGMs).

Since 1970 numerous SRGMs have been developed for the assessment of software reliability. These models incorporate various aspects related to software development for e.g. the code size and complexity, the skill of tester, developer and programmer, testing tools and methodology, resources allocated, etc. A number of researchers and IT practitioners have proposed Non-Homogenous Poisson Process (NHPP) based SRGMs to assess the reliability of the software system (Aggarwal et al., 2019; Anand et al., 2018; Kapur et al., 2011). These models help to predict the number of faults and the time for the next failure on the basis of observed failure data.

These time-dependent SRGMs are divided into two classes: one is perfect debugging where it has been assumed that whenever a new failure occurs the faults causing it is removed immediately and no new faults are introduced in the meantime. The other one is imperfect debugging which further can be split into two types: (a) whenever originally detected faults are removed that do not remove completely this phenomenon is known as imperfect fault removal and (b) it takes several attempts to remove a fault and also some new faults which were previously non-existent may also get generated. This phenomenon re-introduction of faults is known as Error generation (Yamada et al., 1984). In this chapter, we model the failure phenomenon of software incorporating error generation.

Previously it was assumed that faults initiating the failure are removed with complete efficiency but later it was observed that all the encountered faults are not removed i.e. fault removal process is not 100% efficient (Zhang et al., 2003) i.e. only few number of faults are removed out of the overall faults spotted. It can be defined as the ability or effectiveness in the fault removal process. This measure helps the developer to predict the effectiveness of the fault detection and further effort needed. In our study, we have considered Fault Removal Efficiency (FRE) because it is immensely correlated with fault removal process i.e., as the FRE escalates fault removability escalates as well.

Also, it was stated that in practice the number of faults experienced is not the same as the faults removed during the process. Musa (1975) defined Fault Reduction Factor (FRF) as "the ratio of a number of faults removed to the number of failures experienced". This indicates that for the reliability evaluation FRF plays an important role. Experimentally, FRF takes values between zero and one. It has been discovered that FRF could be affected by issues like fault dependency, complex code, human nature, etc. This, in turn, affects the FRF curves which can be increasing, decreasing or constant. Here we consider that FRF follows Weibull distribution while error generation and FRE are considered to be constant. FRF, FRE and error generation all these factors have been incorporated because of their significant impact on the failure process.

In the development phase software engineers also face the challenge to develop a reliable product at the lowest possible cost. In this direction, several release time problems have been formulated to find optimal software release time that incurs a minimum cost of software development to developers (Gautam et al., 2018; Kapur et al., 2012). Based on the failure phenomenon and requirements, developers have to determine the optimal testing time period for which they will perform testing. Infinite testing until all the faults are removed is not feasible for developers since it incurs an enormous cost. On the other hand, inadequate testing may result in catastrophic failures and loss of willingness among the users for the product. To tackle these researchers have devised cost optimization problems that minimize cost along with attaining a level of reliability for the system. In our proposed optimization problem, we consider various costs influencing the development process such as testing and warranty cost, opportunity cost and penalty cost of failures after the warranty expires.

In today's age of competition, companies use product differentiation techniques to mark existence of their product into the market. One of the value-added services that are used widely by the companies is warranty. Software systems are getting complex day by day that creates insecurities among the buyers. In these situations warranty acts as marketing tool. By providing warranty, developer guarantees the user to remove all the faults causing failure during usage for a specified period of time. It benefits the developers by creating a sense of product quality among its users but also incurs some additional cost to developers. Hence they have to maintain balance so that they can gain profits and avoid losses. Market Opportunity cost is dependent on testing time which tends to increase with an increase in testing time. Hence appropriate time to release the software is very necessary. It is defined as loss to the firm due to loss of opportunity caused due to delay in release in monetary terms.

The model is validated using a real fault dataset to obtain optimal release and warranty time. Several methods have been used in the past to solve the software release time optimization problems considering warranty and reliability requirements of the system. One of the most recognized and efficient ways is using soft computing techniques. These techniques became popular because they are simple, flexible, derivation-free and avoids local minima (Mohanty et al., 2010). We have discussed the approach in detail in the future section.

The objective of this study is to use the swarm inspired Metaheuristic Grey Wolf Optimizer (GWO) algorithm to solve the software release time problem. In this study, we obtain optimal release time and optimal warranty time period based on development cost criteria incurred during various phases based on mathematical model that estimates the reliability of software system. The advantage of using GWO over other techniques is that here the optimization problem is continuously controlled by the leader which is not the case of other algorithms. GWO comprises of less number of parameters thus is less cumbersome to deal with. Also this algorithm is versatile in nature because it can be implemented to most of the optimization problems without changing much in the mechanism. Figure 1 presents the research approach followed in this chapter. The main focus of the proposed study is,

- To model the failure process of the software system incorporating FRF, FRE and error generation.
- To examine the model's goodness of fit using real-life software fault dataset.
- To find optimal release time and optimal warranty period based on reliability and cost components using the Grey Wolf optimization algorithm.
- To study the impact of different cost components via sensitivity Analysis on
  - Release time
  - Warranty time period
  - Overall Development Cost
- To identify key cost components affecting the cost.
- To test the significance of error generation and fault removal efficiency on software development cost.

A number of SRGMs and its concerned release planning problems have been proposed in the literature. None of the studies discuss the failure tendency of software systems under the impact of Weibull FRF, FRE and error generation together. Many

Figure 1. Research Methodology

meta-heuristic techniques such as Genetic Algorithm have been used for solving optimization problems in the field of software reliability. But recently some new techniques with better convergence properties have been proposed which have not been employed to solve the software release time problems. Therefore we have made an attempt to utilize the GWO technique to determine the when to release the software into market and for what duration warranty should be given for the software product. This is so because GWO overcomes many limitations of previous techniques.

Due to the impact of different environmental factors on the SRGMs and also in the urge of making high reliable software consisting of all possible functions and features, this situation motivates us to develop a model through this impact and also the different costs involved in developing it. This research is an attempt to provide software engineers an SRGM which can estimates the faults more accurately as it involves different factors that affect the estimation. Also the cost function comprising of opportunity cost, penalty cost and warranty cost, is optimized using GWO. Previously, to obtain the optimal value many soft computing techniques have been used here we have used GWO due to its flexible nature and high performance than the other algorithms.

The study contributes to the literature by using a better optimization technique to solve the software development cost minimization problem. The overall cost of the software system is based on the failure phenomenon modelled by incorporating Weibull FRF, FRE and error generation. The inclusion of practical factors into the models helps in better fault prediction. Therefore incorporating these environmental factors makes the model more accurate and realistic. The proposed study used GWO technique to determine the optimal release and warranty time for the software system. Using of more realistic mathematical model for assessing reliability of software system and one of the latest meta-heuristic techniques GWO for solving optimization problems helps to achieve the best possible results of the release planning problem.

The chapter is organized as follows; in the next section, we have discussed the related concepts and its literature followed by the modelling of SRGM and optimization problem formulation with detailed methodology. Further, we have briefly explained GWO used for the optimization. In the next section application of the proposed approach is illustrated and compared with two standard meta-heuristic techniques along with discussion of the results. Further, we conclude the work and give the limitations and future scope of our study.

BASIC TERMINOLOGIES AND LITERATURE REVIEW

In this section, we will discuss the past and contemporary research being done in the relevant area. Studying literature, in brief, helps to identify the importance of the problem under consideration and find the research gap. All the concepts and terms used are discussed to elaborate and highlight the significance of software release planning by maintaining the reliability under resource constraints. Reliability for a software system is defined as the "probability of failure-free operation for a fixed period of time under the same conditions" (Yamada and Tamura, 2016). This area came more into light when there was a tremendous increase in software demand in almost every. The developers and firms deal with the challenge of providing the best products to end-user.

SRGMs

The reliability of software systems is measured by observing the failure phenomenon. Many SRGMs are based on NHPP assuming the number of failures to be counting process. These models represent the relationship between a number of faults removed or detected with time. The failure process for a general NHPP based SRGMs is modelled by the following differential equations,

$$\frac{dm(t)}{dt} = b(t)\big(a(t) - m(t)\big) \tag{1}$$

Where,

$a(t)$ is fault content,

$b(t)$ is the rate of detecting faults and

$m(t)$ is a measure of a cumulative number of faults detected or removed up to time t.

Solving the differential equation (1), we obtain the expression for Mean Value Function (MVF) $m(t)$. Different expressions are obtained by representing $a(t), b(t)$, as different functions (representing curves) based on the assumptions as well as the dataset. In the past various NHPP based SRGMs have been developed by researchers. One of the very initial SRGM given by Goel andOkumoto (1979) considered finite failure model with constant parameters whose intensity function is exponentially distributed. Many times S-shaped curve for MVF has been observed rather than exponential. This implies an improvement in testing efficiency as the testing progresses with time. This 2-stage model given by Yamada et al. (1984) consists of fault detection and fault isolation phases of the testing process. The model considers delay or time lag between the fault detection and removal assuming that all the detected faults are removed. Kapur andGarg (1992) developed an SRGM assuming that detection of a fault leads to detection of some more faults in the system. They considered that there was some kind of mutual dependence between the faults.

FRF

This factor is defined as "the net number of faults removed is only a proportion of the total number of failures experienced" or as "the net number of faults detected proportional to the number of failures experienced". It was first defined and proposed by Musa (1975). The FRF forms discussed in the literature have been listed in Table 1.

There may be variation in FRF under different situations and environmental factors. There is a noticeable relationship between environmental factors and FRF. Some of the factors influencing FRF are:

- Time difference between detection and removal of faults.
- Imperfection in debugging causes reintroduction or an increase in a number of latent faults. Also sometimes it requires more than one attempt to remove a particular fault. The value of FRF decreases when new faults are introduced.
- Change in efficiency to remove faults due to the introduction of some new faults. Efficiency is affected because all faults are of different severity and require different effort levels to be eliminated.
- Resources allocated for testing and debugging.
- Dependence of fault on some factors or on each other.

Table 2 presents the SRGMs consisting of FRF as the key parameter.

Table 1. Expressions for FRF

References	Explanation	Mathematical Expression for FRF $(0 < F \leq 1)$
Musa (1975), Musa (1980), Musa et al. (1987),	FRF was modelled as the proportion of failures experienced to the number of faults removed.	$F = \dfrac{n}{m}$ m be the number of failures experienced n is the number of faults removed.
Malaiya et al. (1993), Naixin andMalaiya (1996), Musa (1991)	Based on observations the value of FRF can be modelled by fault exposure ratio (at time t).	$F = \dfrac{\lambda_0}{kfm}$ λ_0 : is the initial failure intensity, k : is the faults exposure ratio, f : is the linear execution frequency of the program.
Friedman et al. (1995),	FRF has also defined in the form of detectability, associability and fault growth ratio The Detectability ratio is defined as the number of faults whose resulting failures could be detected. The value of D is near 1, the value of A is near 0 and the value of G lies between 0 and 0.91.	$F = D\big(1 + A\big)\big(1 - G\big),$ D is the detectability ratio, A is the associability ratio, G is the fault growth ratio.

Error Generation

SRGMs formulated assuming that the fault content remains the same throughout is not a practical approach. There is a possibility that faults get generated due to some reasons. This is known as the error generation phenomenon(Goel, 1985; Yamada et al., 1992). The accuracy of the models can be improved by considering the practical issues encountered during fault detection and removal process. This may be the result of several environmental factors that affect the testing process such as tester's skill, code complexity, erroneous code, availability of resources, the severity of faults, etc. The functional form of error generation is considered to be linearly or exponentially increasing with time.

Goel (1985) extended J-M (Jelinski and Moranda, 1972) model considering the imperfect fault removal process. Pham et al. (1999) integrated the imperfect debugging with delayed S-shaped fault detection rate. In later years Kapur and Younes (1996)

Table 2. SRGMs based on FRF

Researchers	FRF $(0 < F \leq 1)$	Model
Musa (1975), Musa (2004)	Musa in his basic Execution time model assumed FRF to be Proportional to the hazard rate function.	$$\frac{dm\left(t\right)}{dt} = Fz\left(t\right) = F\varphi\left[a - m\left(t\right)\right]$$ $z\left(t\right)$ is the hazard rate function φ be the per fault hazard rate. The mean value function is: $$m\left(t\right) = a\left(1 - e^{-F\varphi t}\right)$$
Hsu et al. (2011)	FRF is defined as constant as well as the time variable. Considering that scenario Hsu et. al. defined it to be constant, decreasing and increasing for software with a single release • Constant: $F\left(t\right) = F$ • Decreasing curve: $$F\left(t\right) = F_0 e^{-kt}$$ • Increasing curve: $$F\left(t\right) = 1 - \left(1 - F_0\right)e^{-kt}$$ F_0 represents initial FRF while k is a constant	$$\frac{dm\left(t\right)}{dt} = r\left(t\right)\left(a - m\left(t\right)\right), r\left(t\right) = rF\left(t\right)$$ $m\left(t\right)$ for the three cases is given as: • $m\left(t\right) = a\left(1 - e^{-Frt}\right)$ • $m\left(t\right) = a\left(1 - e^{-\left(\frac{\left(F_0 - 1\right)\left(1 - e^{-kt}\right)}{k} + t\right)}\right)$. • $m\left(t\right) = a\left(1 - e^{-\left(\frac{F_0\left(1 - e^{-kt}\right)}{k}\right)}\right)$. where r is FDR.
Pachauri et al. (2015)	Pachauri et. al. considered Inflexion S-shaped FRF for the multiple versions of the software system under perfect and imperfect debugging environment. In total 3 models were proposed.	$$\frac{dm\left(t\right)}{dt} = rF\left(t\right)\left(a\left(t\right) - m\left(t\right)\right), \ F\left(t\right) = \frac{\alpha}{1 + \beta e^{-\alpha t}}$$ Model1: $m\left(t\right) = a\left(1 - \left(\frac{1 + \beta}{1 + \beta e^{-\alpha t}}\right)^r e^{-\alpha rt}\right)$ Model 2: $\frac{da\left(t\right)}{dt} = \alpha \frac{dm\left(t\right)}{dt}$ $$m\left(t\right) = \frac{a}{1 - \alpha}\left(1 - \left(\frac{1 + \beta}{1 + \beta e^{-\gamma t}}\right)^{r\left(1 - \alpha\right)} e^{-\gamma r\left(1 - \alpha\right)t}\right)$$ Model 3: Extended 2nd model for M-R software. $\alpha : fault\, introdcution\, rate$ $\gamma\, and\, \beta\, are\, shape\, and\, scale\, parameters\, resp.$

continued on following page

Table 2. Continued

Researchers	FRF $(0 < F \leq 1)$	Model
Aggarwal et al. (2017)	Aggarwal et. al. considered the successive releases of the OSS based on Exponentiated Weibull FRF along with the effect of change point and error generation.	$$m_{ir}(t) = \begin{cases} \left(\dfrac{a_i}{(1-\alpha_{i1})} + a_{i-1}^*\right) F_{i1}(t)\tau_i \\ \left(\dfrac{a_i}{(1-\alpha_{i1})} + a_{i-1}^*\right)\left[1 - \dfrac{\left(1-F_{i1}(\tau_i)\right)\left(1-F_{i2}(\tau_i)\right)}{\left(1-F_{i2}(\tau_i)\right)}\right] \\ \dfrac{\alpha_{i1} - \alpha_{i2}}{(1-\alpha_{i1})} F_{i1}(\tau_i) \\ \tau_i T_i \end{cases}$$

introduced the imperfect debugging SRGM considering faults removal process to be exponentially distributed. Jain et al. (2012) proposed the imperfect debugging SRGM with fault reduction factor and multiple change points. Apart from these many other researchers considered imperfect environment conditions of debugging and testing (Chatterjee and Shukla, 2017; Kapur et al., 2011; Li and Pham, 2017a; Sharma et al., 2018; Williams, 2007). Recently many considered FRF factor and error generation together (Anand et al., 2018; Chatterjee and Shukla, 2016; Jain et al., 2012; Kapur et al., 2011)

FRE

During the process of detecting and removing faults, it is not possible to completely remove all the faults latent in the system. This signifies that the testers and debuggers involved in the process are not 100% efficient in removing faults. This efficiency can be measured in percentages. Therefore there was a need to develop a model incorporating a factor representing the efficiency of removing faults. This measure helps the development team to analyze the effectiveness of fault removal process. FRE and faults removed are correlated because higher is the efficiency higher will be the fault removal rate and vice versa. Jones (1996) defined FRE as "percentage of faults removed by review, inspections, and tests." He also briefly described its importance and range for different testing activities. The range is as follows:

1. The efficiency of unit testing ranges from 15% to 50%.
2. The efficiency of integration testing ranges from 25% to 40%.
3. The efficiency of system testing ranges from 25% to 55%.

Liu et al. (2005) proposed an NHPP based SRGM where FRE was defined as the function of debugging time. They assumed the bell-shaped fault detection rate. Later Huang andLyu (2005) considered improvements in testing efficiency with time to give release policy based on reliability. Kremer (1983) proposed the birth-death process in the reliability growth model, assuming imperfect fault removal (death process) and fault generation (birth process). Li andPham (2017b) gave the SRGM model incorporating FRE. He analyzed the accuracy of the model when testing coverage concept is taken together with FRE and error generation parameter.

Release Time Problems

The planning software release is a crucial part of the software development process. The time of release is based on several factors like the failure phenomenon followed by the system, reliability, the cost of testing and debugging during testing, and various kinds of cost the developers have to bear after the release. Three kinds of release problems have been mainly formulated in literature. One is maximizing reliability subject to the resource constraints, second is minimizing development cost subject to reliability constraint and last can be a multi-objective problem with the objective of minimizing cost and maximizing reliability.

Goel andOkumoto (1979) gave a release policy based on reliability and cost criteria. Leung (1992) obtained optimal release time under the constraint of a given budget. Pham (1996) proposed a release planning problem by optimizing cost consisting of penalty cost for failures occurring in the operational phase. Huang andLyu (2005) also proposed a release policy considering improvements in testing efficiency based on reliability and cost criteria. Inoue andYamada (2008) extended existing release policies by incorporating change-point in the SRGM. Li et al. (2010) carried out release time sensitivity analysis by considering a reliability model with testing effort and multiple change points. Kumar et al. (2018) proposed a reliable and cost-effective model considering patching to determine optimal release time.

While planning release policies, firms do not take into account the goodwill loss due to failures after release. This plays a major role in influencing demand. To overcome customer discontent due to failures during the execution of software, the developers offer warranties assuring their users to maintain the software for a time period pre-specified by them. These plans cover all kinds of failures under some specified conditions of usage. This increases the cost burden on firms, hence they need to decide a period length that does not take them into monetary losses. In this direction, Kimura et al. (1999) proposed an optimal software release policy by considering a Warranty policy. Pham and Zhang (1999) proposed a cost optimization model considering warranty cost and risk costs. Dohi (2000) proposed a cost minimization model to determine the optimal period for warranty. Rinsaka andDohi

(2005) also proposed a cost model to minimize the development cost and find an optimal warranty period. These optimization problems consider cost component for providing warranty and cost of removing fault during that period. Kapur et al. (2013) developed a cost model to find optimal release and warranty time considering testing and operational cost. Luo and Wu (2018) optimized warranty policies for different types of failures. They considered that failure could occur due to software failure, hardware failure or due to any human factor.

Software Development is a time-consuming and costly process. Developers aim to achieve desirable reliability, minimizing the cost of development along with satisfying the customer requirements. On the other hand, users demand a reliable product at less cost and good customer service. As already discussed that there are some existing cost models in the literature to challenge the above concern, still, release policies considering warranty plan have not been given much concern. Still, researchers are working to improve the policies due to fast-changing technologies and stipulations from customers

Soft Computing Techniques in Software Reliability

Soft Computing is an optimization technique used to find solutions to real-world problems. These techniques are inspired by nature. It can be categorized into fuzzy logic (FL), Evolutionary, Machine Learning and Probabilistic. This chapter focuses on evolutionary algorithms and more specifically meta-heuristic technique. These techniques are developed based on some animal behavior. With time a number of techniques have been developed like Genetic Algorithm (GA) (Holland, 1992) inspired by the evolutionary mechanism of living beings, Ant Colony Optimization (ACO) (Dorigo and Di Caro, 1999) inspired from the colony of real ants, Particle Swarm Optimization (PSO) (Eberhart and Kennedy, 1995) was inspired from social interactions and many more. Recently also some techniques have been devised to improve the optimization process. These techniques have been widely applied in various areas like engineering, actuarial, process control, etc. The application of soft computing techniques can be extensively seen in the field of hardware reliability for solving various optimization problems (Garg, 2015; Garg and Sharma, 2013). These techniques can be used to solve multi-objective problems (Garg and Sharma, 2013) and for analyzing the performance of industrial systems (Garg, 2017). Recently many models have been developed by integrating two different soft computing techniques to solve optimization problems. Garg (2016) proposed hybrid PSO-GA approach for solving constrained optimization problems. Later, Garg (2019) proposed hybrid (Gravitational Search Algorithm) GSA-GA for solving the constrained optimization problem. In this study, focus is on the applications in the area of software reliability. Table 3 recollects some of the work done in the literature using these techniques.

Table 3. Soft Computing Application in Software reliability

Reference	Technique Used	Explanation
Pai (2006)	Support Vector Machine and GA	Reliability Forecasting
Sheta andAl-Salt (2007)	PSO	Parameter Estimation
Fenton et al. (2008)	Bayesian Network	For predicting software reliability and a failure rate of defects
Kapur et al. (2009)	GA	Testing resource allocation for modular software under cost and reliability constraints
Aljahdali andSheta (2011)	FL	SRGM development using FL
Al-Rahamneh et al. (2011)	Genetic Programming	To develop SRGM
Saed andKadir (2011)	PSO	Performance Prediction
Aggarwal et al. (2012)	GA	Testing Effort Allocation
AL-Saati et al. (2013)	Cuckoo Search	Parameter Estimation
Nasar et al. (2013)	Differential Evolution	Software Testing Effort Allocation
Shanmugam andFlorence (2013)	ACO	To determine the accuracy of software reliability growth models
Tyagi andSharma (2014)	Adaptive Neuro-Fuzzy	To estimate the reliability of a component-based software system.
Kim et al. (2015)	GA	Parameter Estimation
Mao et al. (2015)	ACO	To generate test data for structural testing
Choudhary et al. (2017)	GSA	Parameter Estimation
Choudhary et al. (2018)	Firefly Optimization	Parameter Estimation
Gupta et al. (2018)	Differential Evolution	Test case optimization: Selection and prioritization of test cases for wide-ranging fault detection
Chatterjee andMaji (2018)	Bayesian Network	For software reliability prediction in the early phase of its development
Chaudhary et al. (2019)	Crow Search Optimization	Parameter Estimation
Ahmad andBamnote (2019)	Whale-Crow Optimization	Software Cost Estimation
Proposed Work	GWO	Software Release planning; determining optimal release, warranty time, the overall cost of software development.

After conducting the comprehensive literature survey suggest the gap in using of latest and better techniques for optimizing software development cost and obtain optimal release and warranty time period. The above Table 3 clearly highlights the breach.

PROPOSED MODELLING FRAMEWORK

In this section, we develop a mathematical model for software release planning. The approach can be divided into two steps. In the first step, an NHPP based modelling framework is developed by taking into consideration various factors affecting the failure process and then in the second step cost minimization model is developed by taking into consideration different cost components of testing and operational phase. Release planning of the software is largely affected by the testing conditions and tools; financial and goodwill losses due to failures after release and the warranty cost. The notations used in the subsequent sections are given in Table 4.

Model Assumptions

The proposed SRGM is based on the following assumptions:

1. The failure process is based on general assumptions of a Non-Homogenous Poisson Process.
2. As testing progresses, there is an increase in the fault content of the system at a constant rate. Mathematically it is given as $a(t) = a + \alpha m(t)$.
3. The development team is not 100% efficient in fault removal. Software testing team efficiency is represented by a constant parameter p.
4. Time variable FRF is modelled using Weibull function. Mathematically it is expressed as $F(t) = yt^k$. FRF is modelled using Weibull function because of its property to restore the constant, increasing, decreasing trends.

Based on the above assumptions the mathematical model for Fault Removal Process is developed as follows:

$$\frac{dm(t)}{dt} = b(t)\big(a(t) - pm(t)\big) \tag{2}$$

Where,

Table 4. Notations

t	Time
$m(t)$	Cumulative number of faults detected up to time t
$b(t)$	Rate of Fault Detection
$a(t)$	Fault content at time t
a	Faults Initially present in the software system
α	Constant Rate of Error generation
p	Constant, Fault removal Efficiency
$F(t)$	Time-variable Fault Reduction Factor
y, k	Constant parameters of Weibull function
r	Learning Rate as the testing progresses
C_t	Unit Cost of performing testing
C_{tr}	Per unit cost of removing faults during the testing phase
C_w	Unit Cost of providing Warranty time
C_{wr}	Per unit cost of fault removal during the warranty
C_{opp}	Market Opportunity cost
C_{pw}	Per unit cost of removing faults after the warranty has expired
$Z(t)$	Software development Cost Function
T	Software Release Time
W	: Warranty Time

$$b(t) = F(t)r;\tag{3}$$

$$F(t) = yt^k;\tag{4}$$

$$a(t) = a + \alpha m(t)\tag{5}$$

Using Equations (3), (4) and (5) in Equation (2), we get;

$$\frac{dm(t)}{dt} = yt^k r\left(a + \alpha m(t) - pm(t)\right)\tag{6}$$

Solving the above differential equation using initial condition; at $t = 0; m(t) = 0$, the Mean value function is obtained as follows;

$$m(t) = \frac{a}{p - \alpha}\left[1 - \exp\left(-\frac{yr(p - \alpha)}{k + 1}t^{k+1}\right)\right]\tag{7}$$

Software Development Cost Optimization

Next, we develop the optimization model to minimize the overall cost of software development. Some cost is incurred during every phase of development. In our model, we have considered two phases namely; testing and operational phase of SDLC. The operational phase is further divided into warranty and post-warranty phases (Figure 2).

After the software code has been developed, it is tested for a limited time aiming to remove maximum possible faults from the system before it is released to the market.

Figure 2. Software Product's Life Cycle

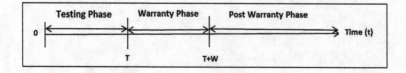

Though the developers strive hard to achieve the maximum reliability of the product through testing but too much testing may not be economically viable for them. Apart from the economic considerations, the developer is also interested in early release to have the 'first entry edge' in the market. There is a strong conflict between the economic and quality aspiration levels. For this optimization, the theory provides valuable tools to look for appropriate release planning while creating a balance between the cost and quality considerations. To differentiate their products in the market the IT firms also provide warranties for support during the post-implementation phase. During warranty, the developer assures the user that any failure / unusual behavior occurring during this time will be fixed by the developer and all the cost will be borne by the developer only. After the warranty expires, if a failure occurs then it leads to goodwill loss of the firm and detrimental to its image. Therefore the developer considers penalty cost for any failure occurring after the warranty expires during the operational phase.

Cost Components

The optimization model considers the costs involved during the two major phases of the development process. These costs components are part of the testing and operational phase.

The testing phase consists of the following three costs:

1. Fixed cost of testing: This cost is incurred due to resources and time spent on testing.
2. Per unit cost of removing faults during the testing phase.
3. Market Opportunity cost

The operational phase consists of three cost components related to Warranty and post Warranty phase:

1. Fixed cost of warranty
2. Per unit cost of dealing with faults encountered during the warranty period.
3. The penalty cost of faults that occur after the warranty expires. This cost is taken into account because these faults result in goodwill loss and may hamper further purchases. Hence the developer needs to control the cost they have to bear due to this.

Cost Model Assumptions

Our proposed software development cost model for release planning is based on the following set of assumptions.

1. Testing (Warranty) cost is linearly related to testing (warranty) time.
2. The cost of removing faults during the testing (warranty) phase is a linear function of the number of faults detected during testing (warranty).
3. FRP is modelled by NHPP based SRGM given by Equation (7).
4. The software cost takes into consideration testing cost; opportunity cost; warranty cost; error removal during the testing phase; error removal during the warranty and after the warranty expires.
5. Any failure post-warranty period may lead to penalty costs to developers.

Cost Model

Considering all the above discussions we obtain total expected software cost and formulate optimization problem with the objective of minimizing the total expected cost of development with two decision variables $T\left(Testing\,Time\right)$ and $W\left(Warranty\,Time\right)$.

$$Min\ Z = C_t T + C_{tr} m\left(T\right) + C_{opp} T^2 + C_w W + C_{wr}\left(m\left(T+W\right) - m\left(T\right)\right)$$
$$+ C_{pw}\left(m\left(\infty\right) - m\left(T+W\right)\right)\left(O_1\right)$$

where

$$T, W > 0$$

Where $m\left(.\right)$ is given by equation (7)

The optimization problem $\left(O_1\right)$ is solved using a soft computing technique called Grey wolf Optimizer (Mirjalili et al., 2014) in MATLAB. In the next section, we discuss the relevance of using this technique for optimization.

Now, we summarize all the steps of the study:

1. Formulate the MVF incorporating the factors affecting the FRP. Here we have considered imperfect debugging due to error generation, the efficiency of the testing team and FRF

2.   Estimation of SRGM parameters using a real-life fault dataset of web-based ERP systems.

3.   Determine cost components for the optimization model based on the life cycle of a software product.

4.   Formulate the cost minimization problem with two decision variables T and W i.e. release time and warranty time period.

5.   Apply GWO in MATLAB to obtain the optimal values for T and W and then evaluate the costs.

6.   To determine the key cost component, sensitivity analysis is done on each cost component. This helps to visualize the impact of the increase/decrease in cost by 10% on the release time, warranty time and total cost.

GREY WOLF OPTIMIZER

Metaheuristics represent "higher level" heuristic-based soft computing algorithms that can be directed towards a variety of different optimization problems by instantiating generic schema to individual problems, needing relatively few modifications to be made in each specific case. Meta-heuristic techniques have become popular due to four major reasons. Firstly it's easy to understand the concepts inspired by natural phenomena. Secondly, it can be applied to various problems without actually making modifications to the algorithm. These algorithms solve the problem as black boxes. Thirdly these techniques are stochastically computed by starting optimization with a random solution, without having to calculate derivatives of each search space. Hence they are derivative-free. Lastly, it avoids the local optimal solution. This is probably due to their stochastic nature, which helps them to move from local to global solutions.

GWO (Mirjalili et al., 2014) is a meta-heuristic algorithm inspired by the social hierarchy and hunting behavior of Grey Wolves. The solution convergence of this technique explores search space as a multi-level decision mechanism and does not require gradient for the search path. GWO is versatile for optimization problems because of its easy implementation process and few parameters in the algorithm. This is a powerful method to avoid premature convergence by expanding the search area and speeds up the optimization process. This has been applied in quite a few areas like economic dispatch problems (Jayabarathi et al., 2016), parameter estimation (Song et al., 2015), etc. This method is better in convergence than other methods because of following advantages:

1.   It has a good procedure for conveying and sharing the information.

2. It considers three solutions (α, β, δ) to get the best optimization results considering a random function. This helps to fasten the process of moving from local to global optimal values.

3. There are only two main parameters that need to be adjusted for the transition between exploration and exploitation.

Grey Wolves follow a very strict social leadership hierarchy and policy for chasing prey. These are generally found in packs consisting of 5-12 wolves. Further, they are divided into four levels of hierarchy (Figure 3). Level one i.e. the one at the top of the hierarchy corresponds to the pack leader alpha (α), which can be either male or female. All the important decisions like the sleeping place, hunting, etc. are taken by alphas and dictated to other members. The power and role of the grey wolf decrease as we move down the hierarchy. The next level is of subordinates that help leaders in making decisions. These are known as beta (β). Betas help to maintain discipline and implement the decisions taken by the leaders. In case alpha gets old or dies untimely, beta takes its place. Delta (δ) dominates the omegas (ω) at the lowest level and report to the betas. Omegas are allowed to eat.

This hierarchy also represents the level of solutions i.e. alpha (α) is considered to be the best solution, beta (β) is the second-best solution and deltas (δ) is the third-best solution. The rest of the solutions are leftover and taken as omegas (ω). The alpha, beta and delta guide the optimization process while omegas follow them.

Figure 3. Social Hierarchy of Grey Wolves (In Decreasing order of Dominance)

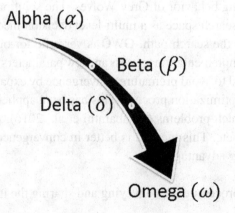

Alpha (α)

Beta (β)

Delta (δ)

Omega (ω)

Grey wolves are not only social because of their hierarchy but their group hunting is also a special interesting social behavior. Broadly the whole process can be categorized into two phases:

1. Exploration: Here the possible areas where the prey could be located are identified.
2. Exploitation: The areas identified are searched thoroughly and attacking is done.

The hunting process of grey wolves is divided into the following steps:

1. Search for prey (Exploration)
2. Approaching the prey by tracking and chasing
3. Encircling the prey
4. Harassing the prey till it stops its movement
5. Attaching prey (Exploitation)

The mathematical model and algorithm have been broadly explained by Mirjalili et al. (2014). The whole process followed by the GWO algorithm is depicted through a flow chart (Figure 4).

We will discuss the step by step process thoroughly.

1. Initialization of GWO parameters such as maximum number of Iterations, Population size, number of search agents and vectors $d, A \, and \, C$. These vectors are initialized using a random function. (Equations 8 and 9)

$$\vec{A} = 2\vec{d}.rand_1 - \vec{d} \tag{8}$$

$$\vec{C} = 2.rand_2 \tag{9}$$

Where d linearly decreases from 2 to 0 as iteration proceeds

2. Random generation of wolves based on their pack sizes.
3. The fitness value of each search agent is calculated using the below-mentioned equations;

Figure 4. Flowchart for Optimization using Grey Wolf Optimizer

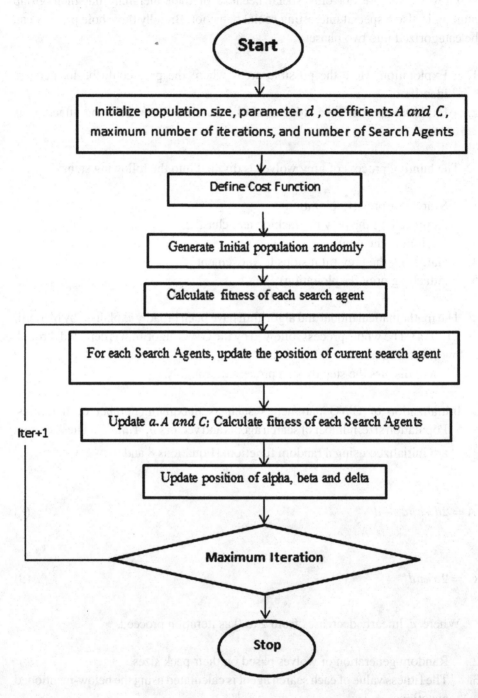

22

$$\vec{D} = \left| \vec{C}.\vec{H}_p(t) - \vec{H}(t) \right| \tag{10}$$

$$\vec{H}(t+1) = \left| \vec{H}_p(t) - \vec{A}.\vec{D} \right| \tag{11}$$

Where, $\vec{H}_p(t)$ is the position of prey and $\vec{H}(t)$ is the wolf position. Equation (10) calculates the distance between the prey and wolf and then accordingly the wolf updates his/her position using equation (11).

4.   Obtain the best search agent (α), the second-best one (β) and the third-best search agent (δ) using equations (11-16);

$$\overrightarrow{D_\alpha} = \left| \overrightarrow{C_1}.\overrightarrow{H_\alpha} - \vec{H} \right| \tag{12}$$

$$\overrightarrow{D_\beta} = \left| \overrightarrow{C_1}.\overrightarrow{H_\beta} - \vec{H} \right| \tag{13}$$

$$\overrightarrow{D_\delta} = \left| \overrightarrow{C_1}.\overrightarrow{H_\delta} - \vec{H} \right| \tag{14}$$

$$\overrightarrow{H_1} = \vec{H}_\alpha - \overrightarrow{A_1}.(\overrightarrow{D_\alpha}) \tag{15}$$

$$\overrightarrow{H_2} = \vec{H}_\beta - \overrightarrow{A_2}.(\overrightarrow{D_\beta}) \tag{16}$$

$$\overrightarrow{H_3} = \vec{H}_\delta - \overrightarrow{A_3}.(\overrightarrow{D_\delta}) \tag{17}$$

5.   Based on the locations of three beast search agent the location of the current one is updated using equation (18);

$$\vec{H}(t+1) = \frac{\overrightarrow{H_1} + \overrightarrow{H_2} + \overrightarrow{H_3}}{3} \qquad (18)$$

6. Again the fitness values of all the agents are calculated
7. The values of $\vec{H}_\alpha, \vec{H}_\beta$, and $\vec{H}_\delta$ are updated accordingly.
8. When stopping condition reaches print the solution, otherwise repeat step 5 for all the iterations

GWO algorithm consists of a few parameters, but they are a very important role in convergence. Hence we discuss the significance of these parameters in detail. The grey wolves diverge to search for prey (exploration) and converge to attack (exploitation). Mathematically this is achieved through the parameter $\vec{A}$. Its value is influenced by $\vec{a}$. To model a wolf approaching towards the prey, the value of $\vec{a}$ is decreased up to zero from two. If $|A| > 1$ then the wolf moves away from the prey in hope to find a better prey otherwise if $|A| < 1$ then wolf moves towards the prey for attacking.

Component $\vec{C}$ also helps in exploration. It provides random weights to acknowledge the impact of prey in defining the distance. If $C > 1$ then it emphasizes its effect otherwise it deemphasizes the effect of prey. C is deliberately decreased randomly throughout the process from first to last iterations which are definitely not a linear decrease. Thus it is helpful in bringing the local solutions to halt. On studying these parameters thoroughly we can sum the process of searching the prey into steps as:

1. The random population of wolves (candidate solutions) is generated.
2. The wolves at the top three levels of hierarchy determine the position of prey.
3. All the candidates update their position from prey.
4. $\vec{a}$ decreased from 2 to 0 to emphasize exploration and exploitation.
5. $\vec{A}$ value tends the wolves to either converge or diverge form prey.
6. Hence after satisfaction, the algorithm is terminated.

NUMERICAL RESULTS AND DISCUSSIONS

Data Description

The proposed SRGM needs to be validated on a real-life fault dataset before employing it for optimal release planning. For this, we have used the fault data set obtained from web-based integrated ERP systems. The data was collected from August 2003 to July 2008 (SourceForge.net, 2008). The data has been used in literature to validate the MVF (Hsu et al., 2011; Li and Pham, 2017a). The data set consists of 146 faults observed in 60 months. 60 observations in a dataset to validate the model is quite a large time series. Before estimating the parameters of the model we need to ensure that the dataset is appropriate for the considered model. Hence we have applied *Laplace Trend* analysis. This helps to analyze the trend depicted by the dataset, which can be increasing, decreasing or stable. This test was given and described by Ascher andFeingold (1978). Later it was extended by Ascher andFeingold (1984) and Kanoun (1989).

The method calculates the Laplace factor for each time unit as the testing progress. The value of the factor including the sign determines the trend. Negative Laplace factor implies growth in reliability, positive one implies decay whereas if the Laplace factor value is between $-2\ to + 2$ and it is stable. For the dataset considered in our study, we observed the decay in reliability at the end of 60 months of testing. The Laplace factor value obtained using equation (19) is 4.444 which implies that there is decay in reliability.

$$k\left(t\right) = \frac{\sum_{i=1}^{t}\left(i-1\right)h\left(i\right) - \dfrac{t-1}{2}\sum_{i=1}^{t}h\left(i\right)}{\sqrt{\dfrac{t^2-1}{12}\sum_{i=1}^{t}h\left(i\right)}} \tag{19}$$

The decay trend (Figure 5) shown by the data can only be encountered using a distribution that can capture the S-Shaped failure phenomenon. Hence the proposed fault removal process is appropriate for the fault dataset as it depicts the S-Shaped failure curve.

Parameter Estimation and SRGM Validation

Using the above-mentioned dataset, we estimate the unknown parameters of the model in SPSS using the LSE method. The estimates are given in Table 5.

The goodness of fit curve (Figure 6) shows that the model has a good fit to the data. The values of performance criteria (Table 6) evaluated goes well with the values reported in the literature. The table compares the obtained results with the results presented in Li andPham (2017a) The various performance measures used are the Coefficient of determination (R^2) which is the outcome of regression analysis that shows how much variance of the dependent variable is being explained by the independent variables and Mean square error (MSE).

Figure 5. Reliability Growth Trend depicted by data

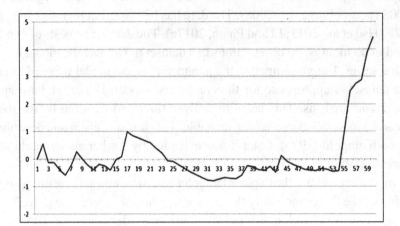

Boxplot

Boxplot is a graphic representation of spread in the data by dividing it into quartiles. It is also termed as whiskers plot since whiskers are present at upper and lower ends of the box representing the top and bottom 25% of values respectively. The last data point at the lower end of the whisker is the minimum value while the topmost value at the whisker is the maximum value. This data excludes the outliers that may be present in the data which are plotted separately as the single data points. The interquartile range i.e. the area covered by the box consists of 50% data with median lying within the box depicted using a line. In this study, we have plotted the boxplots for absolute errors between the observed faults during testing and the predicted faults using the proposed model for each calendar time

Table 5. Estimated Parameters

Parameter	Estimated Value
a	147
α	0.623
p	0.377
k	0.096
y	0.196
r	0.054

Figure 6. Goodness-of-fit Curve

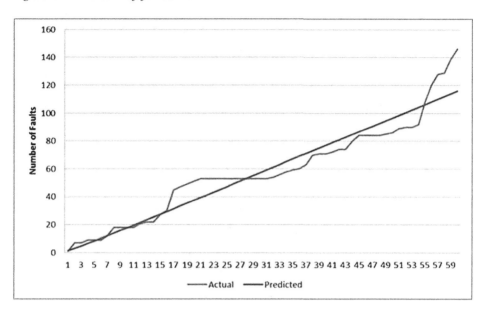

From Figure 7 we can observe that most of the data points lie at the lower end of the boxplot forming a short box implying a good fit. This suggests that the difference between the observed and predicted faults is less here.

Optimization Using GWO

Now we do release planning for the software system considering the optimization problem O_1 discussed in the previous section. Firstly need to assign values to all the unknown parameters. The parameters of MVF are replaced with values obtained

Table 6. Model Performance Analysis

Model	R^2	MSE	PP	PRR
Li andPham (2017a)	0.934	85.955	5.0095	4.428
Chang et al. (2014)	0.927	94.451	2.898	1.795
Pham (2014)	0.883	151.356	2.491	4.827
Roy et al. (2014)	0.928	90.462	2.274	2.213
Zhang et al. (2003)	0.922	101.947	$4.99e^{+4}$	5.093
Proposed Model	0.927	86.627	1.151	1.374

Figure 7. Boxplot for ERP software

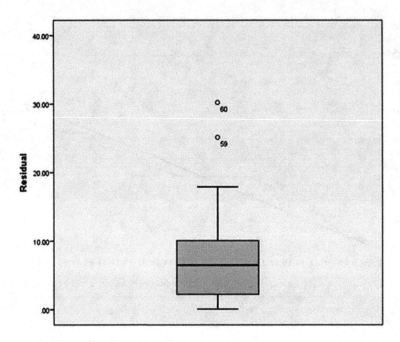

from estimation (Table 4) and cost components are assigned values based on the literature (Kapur et al., 2013; Pham and Zhang, 1999).

Considering the cost components values as:

$$C_t = 2,\ C_{tr} = 3,\ C_w = 2.5,\ C_{rw} = 2.75,\ C_{opp} = 0.5, C_{pw} = 5$$

The optimization problem to be solved using GWO in MATLAB is

$$MinZ = 2T + 3m(T) + 0.5T^2 + 2.5W + 2.75\Big(m(T+W) - m(T)\Big)$$
$$+5\Big(m(\infty) - m(T+W)\Big)(O_2)$$

where

$$T, W > 0$$

Where $m(t)$ is given by equation (7).

The parameters of the mathematical model have been estimated in SPSS. The cost parameters of the software cost model have been assumed based on the literature and industrial practices. The number of search agents and maximum number of iterations required for carrying out GWO was assumed on the basis of previous studies. The number of search Agents was initialized to be 30 and the maximum number of iterations was set to be 150. Alpha, Beta and Delta solutions were initialized with zeros based on the number of variables in the optimization problem. The vector d is initialized using maximum number of iteration given by equation (20) and it tends to decrease from 2 to 0. The vector $A's$ are initialized through d and random variable between 0 and 1 (Equation 8). The vector $C's$ are generated through random variable between 0 and 1 (Equation 9).

$$d = 2 - 1 * ((2)/\text{Max\_iter}) \tag{20}$$

The results obtained are given in Table 7. Figure 8 shows the convergence curve followed by GWO for solving the optimization problem O_2.

From the results we can observe that:

1. Cost components of the testing phase are the major part of the total cost.
2. Timely release of the software is very important otherwise it increases the loss due to the increase in market opportunity cost.
3. The penalty cost due to failures in the post-warranty phase must be controlled by properly analyzing the software during testing.
4. Providing a warranty helps to reduce the penalty cost by taking responsibility for early failures. If the developer does not give warranty cover for its product then any failure after release adds up to penalty cost. This leads to goodwill loss among the users.

Table 7. GWO Results

Optimal Warranty Time Period		12 Months
Optimal Software Release Time		64 Months
Cost ($)		
Testing Phase Cost	Testing Cost	788
	Opportunity Cost	1352
Operational Phase Cost	Warranty Cost	188
	Post-Warranty Cost	1936
Total Cost		4258

To establish stability and feasibility of GWO, its performance has been compared with standard and well-known GA and PSO. GA and PSO are very commonly used and widely accepted algorithm in Software Reliability Engineering (Al-Rahamneh et al., 2011; Saed and Kadir, 2011; Srivastava and Kim, 2009; Windisch et al., 2007). Optimization results are given in Table 8. The results show that GWO gives better fitness value (Minimum cost of development) for optimization. In future we may use other meta-heuristic techniques for doing optimization.

Cost Sensitivity Analysis

Next, we study the effect of each cost component on the overall cost of software development. Sensitivity analysis is a very helpful way to understand the importance of cost components. It studies the impact of the independent variable on the dependent variable under the same conditions. It helps the developers to decide about the cost that they to focus on most since it has maximum influence on the overall cost. Hence it helps to identify the key inputs that extensively affect the outcome.

In our study, we have studied the effect of a change in each cost component on the optimal release time of the software, warranty time period and the corresponding total cost of software development. Firstly each cost factor is increased by 10% then decreased by 10% and observe the relative change. Relative Change can be defined as "determine the percentage change in the dependent quantity with a change in an independent variable". Mathematically is expressed as:

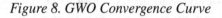

Figure 8. GWO Convergence Curve

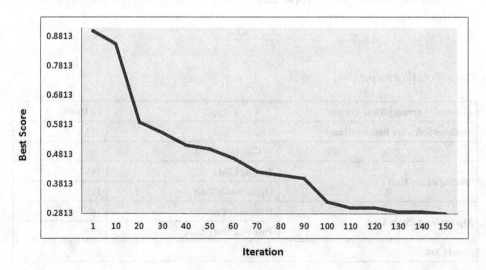

Table 8. Comparison Results

Algorithm	Optimal Testing Time	Optimal Warranty Time	Optimal Development Cost
GWO	64	12	4258
PSO	60	24	4526
GA	60	30	4518

$$Relative\, Change = \frac{new\, value - old\, value}{old\, value} \qquad (21)$$

Tabular representation of the values obtained on sensitivity analysis is given in Table 9. Figures 9 and 10 give the graphical view of relative changes in release time, warranty time and cost on increase and decrease respectively.

Following observations can be made from Table 9:

1. An increase in testing removal cost persuades the developers to reduce testing time and provide warranty cover for the larger period and vice versa. Hence the testing duration decreases and warranty duration increases with increases in testing cost.
2. An increase in market opportunity costs leads to the early release of the software with an increased warranty time period. Releasing quickly forces the developers to provide a warranty for more time than usual.
3. Also, it is observed that an increase in warranty costs leads to a maximum relative change in the warranty period. In such cases, the developer spends more time on testing to reduce the risk of faults after release and hence control the cost due to warranty.
4. Whereas the increase in penalty cost leads to longer warranty and testing time periods allotted for the software product and vice-versa. Developers can't afford losses due to failures during the operational phase. So to avoid the situation firms test the software for longer time periods and provide a bit larger warranty coverage.
5. Testing Cost is the key cost component that majorly affects the overall cost. The developer needs to focus on this particular component to control the total cost incurred.

Table 9. Cost Sensitivity Analysis

Cost Factor	Cost		Warranty Time (W^*) (Months)	Relative Change in Warranty Time	Testing time (T^*) (Months)	Relative Change in Testing Time	Relative Change in Total Cost
	Original Cost	Modified Cost					
Increase by 10%							
C_t	2	2.2	14	0.166667	62	-0.03125	0.009629
C_{opp}	0.5	0.55	13	0.083333	62	-0.03125	0.03241
C_{rt}	3	3.3	15	0.25	61	-0.04688	0.00775
C_w	2.5	2.75	10	-0.16667	67	0.046875	0.002114
C_{rw}	2.75	3.025	9	-0.25	68	0.0625	0.00047
C_{pw}	5	5.5	13	0.083333	65	0.015625	0.042978
Decrease by 10%							
C_t	2	1.8	11	-0.08333	67	0.047619	-0.01174
C_{opp}	0.5	0.45	11	-0.08333	65	0.015873	-0.03171
C_{rt}	3	2.7	10	-0.16667	68	0.063492	-0.0108
C_w	2.5	2.25	15	0.25	63	-0.01587	-0.00493
C_{rw}	2.75	2.47	16	0.333333	62	-0.03175	-0.00423
C_{pw}	5	4.5	11	-0.08333	63	-0.01587	-0.04415

Impact of Parameters on Total Cost

In order to study the effect of error generation (α) on total development cost, we have calculated the total development cost when the error generation parameter is decreased by 10% each time. The gradual decrease results in a decrease in overall cost. Similarly, by increasing the testing efficiency (p) by 10%, we see a decrease

Figure 9. Relative Change on 10% Increase in Cost Component

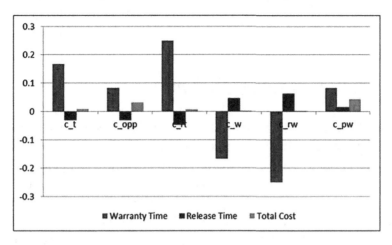

in the total cost. The observations are very clear from Figure 11. If the developer is able to control these parameters then it is also possible to control the cost. Figure 11 provides two important observations:

1. that even a small decrease in the rate of error generation, makes the cost to fall drastically. This is because if the fewer errors are generated and the faults are removed with the same rate then the same effort is consumed to remove a fewer number of faults. Hence there is a sudden decrease in the cost.

Figure 10. Relative Change on 10% Decrease in Cost Component

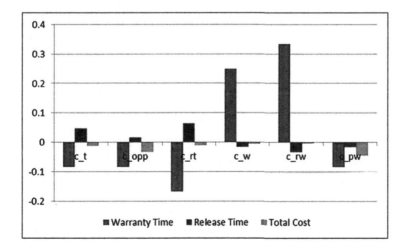

2. On the other hand, if testing capabilities improve but error generation also increases then decrease is cost is less as compared to the previous case.

CONCLUSION AND FUTURE SCOPE

Due to the size and complexities of software systems, detection and correction of faults are challenging and costly. Even after the release of the software, penalty cost is increased and finally loss of goodwill. Thus during the testing phase engineers focus on the maximum removal of the faults to develop a reliable product. Thus it is important to discuss a trade-off between the release time and the reliability of software before release. Here in this chapter, we have discussed the optimal release planning by using GWO. We have also discussed the cost sensitivity analysis and showed the importance of the cost parameters in the software development process.

Initially, we developed an SRGM considering the Error Generation, Fault Removal Efficiency and Fault Reduction Factor, where FRF follows Weibull distribution. For the validation of our model, we have used a real-life dataset of faults from a web-based ERP system and applied Laplace Trend Analysis to check if the data is appropriate for the proposed SRGM. In our study, Laplace factor is 4.444 which is positive in sign and thus indicates decay in the reliability. We have considered the

Figure 11. Impact of error generation parameter and FRE

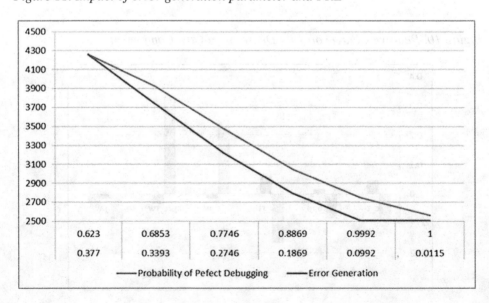

s-shaped model that can fit any dataset following any trend and hence it can capture the failure occurrence of the data with decay trend.

Now we have estimated the parameters of the SRGM using non-linear regression and to analyze its performance we have used different performance measures such as Mean Square Error, Predictive Power and Predictive Ratio Risk. We calculated the Coefficient of determination that gives the variance explained by the independent variable. The goodness of the fit curve is also being plotted which indicates good predictive power of the model.

Once the model is validated release planning problem is discussed for the software system. For this, we formulated a cost optimization function involving testing cost, warranty cost, penalty cost, opportunity cost with two decision variables testing time and warranty time. We solved the problem using the GWO technique to obtain the optimal release time and optimal warranty time by minimizing the total expected cost and the results obtained are shown in Table 7. The results of GWO was compared to some well known meta-heuristic techniques to show its stability and feasibility with the existing models. We also carried out a cost sensitivity analysis to analyze different cost parameters (Table 9). Through the cost sensitivity, we have discussed the cost which has a large impact on the overall cost.

Our study is significant for assessing the reliability and optimizing the release time and warranty with the help of GWO, a meta-heuristic technique. It assists the developer in the testing process with detecting and removing faults while some factors impacting the testing. it also helps the developer to focus on the cost influencing the overall cost.

LIMITATIONS AND FUTURE SCOPE

The limitations of the current study can be overcome by working on the following perspectives of the proposed study:

- We may integrate the model using a multi-release context as it analyses the release time for a single release of the software.
- In the given chapter we carried out two factors FRF, FRE that control failure process but this research can further be extended with other factors such as testing coverage, resource consumptions.
- Instead of using GWO to obtain the optimal release time and warranty time, we can also use other meta-heuristic approaches and can make a comparison among the results.
- We have validated our proposed model using a single dataset but in future more datasets can also be included in the discussion.

- The proposed release problem involves warranty cost, penalty cost and opportunity cost whereas other costs influencing the total cost may be included in the objective function.
- Here we have taken a single objective function of cost minimization but we can also consider the problem when there is a multi-objective optimization function.

REFERENCES

Aggarwal, A. G., Dhaka, V., & Nijhawan, N. (2017). Reliability analysis for multi-release open-source software systems with change point and exponentiated Weibull fault reduction factor. *Life Cycle Reliability and Safety Engineering*, 6(1), 3–14. doi:10.100741872-017-0001-0

Aggarwal, A. G., Gandhi, N., Verma, V., & Tandon, A. (2019). Multi-Release Software Reliability Growth Assessment: An approach incorporating Fault Reduction Factor and Imperfect Debugging. *International Journal of Mathematics in Operational Research.*, 15(4), 446. doi:10.1504/IJMOR.2019.103006

Aggarwal, A. G., Kapur, P., Kaur, G., & Kumar, R. (2012). Genetic algorithm based optimal testing effort allocation problem for modular software. *Bharati Vidyapeeth's Institute of Computer Applications and Management (BVICAM)*, 445.

Ahmad, S. W., & Bamnote, G. (2019). Whale–crow optimization (WCO)-based Optimal Regression model for Software Cost Estimation. *Sadhana*, 44(4), 94. doi:10.100712046-019-1085-1

Al-Rahamneh, Z., Reyalat, M., Sheta, A. F., Bani-Ahmad, S., & Al-Oqeili, S. (2011). A new software reliability growth model: Genetic-programming-based approach. *Journal of Software Engineering and Applications*, 4(8), 476–481. doi:10.4236/jsea.2011.48054

AL-Saati, D., Akram, N., & Abd-AlKareem, M. (2013). *The use of cuckoo search in estimating the parameters of software reliability growth models.* arXiv preprint arXiv:1307.6023

Aljahdali, S., & Sheta, A. F. (2011). *Predicting the reliability of software systems using fuzzy logic.* Paper presented at the 2011 Eighth International Conference on Information Technology: New Generations. 10.1109/ITNG.2011.14

Anand, S., Verma, V., & Aggarwal, A. G. (2018). 2-Dimensional Multi-Release Software Reliability Modelling considering Fault Reduction Factor under imperfect debugging. *INGENIERIA SOLIDARIA, 14.*

Ascher, H., & Feingold, H. (1978). *Application of Laplace's test to repairable system reliability.* Paper presented at the Actes du 1er Colloque International de Fiabilité et de Maintenabilité.

Ascher, H., & Feingold, H. (1984). *Repairable systems reliability: modeling, inference, misconceptions and their causes.* M. Dekker.

Chanda, U., Tandon, A., & Kapur, P. K. (2010). Software Release Policy based on Change Point Considering Risk Cost. *Advances in Information Theory and Operations Research*, 111-122.

Chang, I. H., Pham, H., Lee, S. W., & Song, K. Y. (2014). A testing-coverage software reliability model with the uncertainty of operating environments. *International Journal of Systems Science: Operations & Logistics*, *1*(4), 220–227.

Chatterjee, S., & Maji, B. (2018). A bayesian belief network based model for predicting software faults in early phase of software development process. *Applied Intelligence*, 1–15.

Chatterjee, S., & Shukla, A. (2016). Modeling and analysis of software fault detection and correction process through weibull-type fault reduction factor, change point and imperfect debugging. *Arabian Journal for Science and Engineering*, *41*(12), 5009–5025. doi:10.100713369-016-2189-0

Chatterjee, S., & Shukla, A. (2017). An ideal software release policy for an improved software reliability growth model incorporating imperfect debugging with fault removal efficiency and change point. *Asia-Pacific Journal of Operational Research*, *34*(03), 1740017. doi:10.1142/S0217595917400176

Chaudhary, A., Agarwal, A. P., Rana, A., & Kumar, V. (2019). *Crow Search Optimization Based Approach for Parameter Estimation of SRGMs.* Paper presented at the 2019 Amity International Conference on Artificial Intelligence (AICAI). 10.1109/AICAI.2019.8701318

Choudhary, A., Baghel, A. S., & Sangwan, O. P. (2017). An efficient parameter estimation of software reliability growth models using gravitational search algorithm. *International Journal of System Assurance Engineering and Management*, *8*(1), 79–88. doi:10.100713198-016-0541-0

Choudhary, A., Baghel, A. S., & Sangwan, O. P. (2018). *Parameter Estimation of Software Reliability Model Using Firefly Optimization. In Data Engineering and Intelligent Computing* (pp. 407–415). Springer.

Dohi, T. (2000). The age-dependent optimal warranty policy and its application to software maintenance contract. *Proceedings of 5th International Conference on Probabilistic Safety Assessment and Management*.

Dorigo, M., & Di Caro, G. (1999). Ant colony optimization: a new meta-heuristic. *Proceedings of the 1999 congress on evolutionary computation-CEC99*. 10.1109/CEC.1999.782657

Eberhart, R., & Kennedy, J. (1995). A new optimizer using particle swarm theory. *Proceedings of the Sixth International Symposium on Micro Machine and Human Science*. 10.1109/MHS.1995.494215

Fenton, N., Neil, M., & Marquez, D. (2008). Using Bayesian networks to predict software defects and reliability. *Proceedings of the Institution of Mechanical Engineers, Part O: Journal of Risk and Reliability, 222*(4), 701–712.

Friedman, M. A., Tran, P. Y., & Goddard, P. I. (1995). *Reliability of software intensive systems*. William Andrew.

Garg, H. (2015). An efficient biogeography based optimization algorithm for solving reliability optimization problems. *Swarm and Evolutionary Computation, 24*, 1–10. doi:10.1016/j.swevo.2015.05.001

Garg, H. (2016). A hybrid PSO-GA algorithm for constrained optimization problems. *Applied Mathematics and Computation, 274*, 292–305. doi:10.1016/j.amc.2015.11.001

Garg, H. (2017). Performance analysis of an industrial system using soft computing based hybridized technique. *Journal of the Brazilian Society of Mechanical Sciences and Engineering, 39*(4), 1441–1451. doi:10.100740430-016-0552-4

Garg, H. (2019). A hybrid GSA-GA algorithm for constrained optimization problems. *Information Sciences, 478*, 499–523. doi:10.1016/j.ins.2018.11.041

Garg, H., & Sharma, S. (2013). Multi-objective reliability-redundancy allocation problem using particle swarm optimization. *Computers & Industrial Engineering, 64*(1), 247–255. doi:10.1016/j.cie.2012.09.015

Gautam, S., Kumar, D., & Patnaik, L. (2018). *Selection of Optimal Method of Software Release Time Incorporating Imperfect Debugging*. Paper presented at the 4th International Conference on Computational Intelligence & Communication Technology (CICT). 10.1109/CIACT.2018.8480133

Goel, A. L. (1985). Software reliability models: Assumptions, limitations, and applicability. *IEEE Transactions on Software Engineering, SE-11*(12), 1411–1423. doi:10.1109/TSE.1985.232177

Goel, A. L., & Okumoto, K. (1979). Time-dependent error-detection rate model for software reliability and other performance measures. *IEEE Transactions on Reliability, 28*(3), 206–211. doi:10.1109/TR.1979.5220566

Gupta, V., Singh, A., Sharma, K., & Mittal, H. (2018). *A novel differential evolution test case optimisation (detco) technique for branch coverage fault detection. In Smart Computing and Informatics* (pp. 245–254). Springer.

Holland, J. H. (1992). Genetic algorithms. *Scientific American, 267*(1), 66–73. doi:10.1038cientificamerican0792-66

Hsu, C.-J., Huang, C.-Y., & Chang, J.-R. (2011). Enhancing software reliability modeling and prediction through the introduction of time-variable fault reduction factor. *Applied Mathematical Modelling, 35*(1), 506–521. doi:10.1016/j.apm.2010.07.017

Huang, C.-Y., & Lyu, M. R. (2005). Optimal release time for software systems considering cost, testing-effort, and test efficiency. *IEEE Transactions on Reliability, 54*(4), 583–591. doi:10.1109/TR.2005.859230

Inoue, S., & Yamada, S. (2008). *Optimal software release policy with change-point*. Paper presented at the International Conference on Industrial Engineering and Engineering Management, Singapore. 10.1109/IEEM.2008.4737925

Jain, M., Manjula, T., & Gulati, T. (2012). Software reliability growth model (SRGM) with imperfect debugging, fault reduction factor and multiple change-point. *Proceedings of the International Conference on Soft Computing for Problem Solving (SocProS 2011)*.

Jayabarathi, T., Raghunathan, T., Adarsh, B., & Suganthan, P. N. (2016). Economic dispatch using hybrid grey wolf optimizer. *Energy, 111*, 630–641. doi:10.1016/j.energy.2016.05.105

Jelinski, Z., & Moranda, P. (1972). *Software reliability research. In Statistical computer performance evaluation* (pp. 465–484). Elsevier. doi:10.1016/B978-0-12-266950-7.50028-1

Jones, C. (1996). Software defect-removal efficiency. *Computer, 29*(4), 94–95. doi:10.1109/2.488361

Kanoun, K. (1989). *Software dependability growth: characterization, modeling, evaluation.* Doctorat ès-Sciences thesis, Institut National polytechnique de Toulouse, LAAS report (89-320).

Kapur, P., Aggarwal, A. G., Kapoor, K., & Kaur, G. (2009). Optimal testing resource allocation for modular software considering cost, testing effort and reliability using genetic algorithm. *International Journal of Reliability Quality and Safety Engineering, 16*(06), 495–508. doi:10.1142/S0218539309003538

Kapur, P., & Garg, R. (1992). A software reliability growth model for an error-removal phenomenon. *Software Engineering Journal, 7*(4), 291–294. doi:10.1049ej.1992.0030

Kapur, P., Pham, H., Aggarwal, A. G., & Kaur, G. (2012). Two dimensional multi-release software reliability modeling and optimal release planning. *IEEE Transactions on Reliability, 61*(3), 758–768. doi:10.1109/TR.2012.2207531

Kapur, P., Pham, H., Anand, S., & Yadav, K. (2011). A unified approach for developing software reliability growth models in the presence of imperfect debugging and error generation. *IEEE Transactions on Reliability, 60*(1), 331–340. doi:10.1109/TR.2010.2103590

Kapur, P., & Younes, S. (1996). Modelling an imperfect debugging phenomenon in software reliability. *Microelectronics and Reliability, 36*(5), 645–650. doi:10.1016/0026-2714(95)00157-3

Kapur, P. K., Yamada, S., Aggarwal, A. G., & Shrivastava, A. K. (2013). Optimal price and release time of a software under warranty. *International Journal of Reliability Quality and Safety Engineering, 20*(03), 1340004. doi:10.1142/S0218539313400044

Kim, T., Lee, K., & Baik, J. (2015). An effective approach to estimating the parameters of software reliability growth models using a real-valued genetic algorithm. *Journal of Systems and Software, 102,* 134–144. doi:10.1016/j.jss.2015.01.001

Kimura, M., Toyota, T., & Yamada, S. (1999). Economic analysis of software release problems with warranty cost and reliability requirement. *Reliability Engineering & System Safety, 66*(1), 49–55. doi:10.1016/S0951-8320(99)00020-4

Kremer, W. (1983). Birth-death and bug counting. *IEEE Transactions on Reliability*, *32*(1), 37–47. doi:10.1109/TR.1983.5221472

Kumar, V., Singh, V., Dhamija, A., & Srivastav, S. (2018). Cost-Reliability-Optimal Release Time of Software with Patching Considered. *International Journal of Reliability Quality and Safety Engineering*, *25*(04), 1850018. doi:10.1142/S0218539318500183

Leung, Y.-W. (1992). Optimum software release time with a given cost budget. *Journal of Systems and Software*, *17*(3), 233–242. doi:10.1016/0164-1212(92)90112-W

Li, Q., & Pham, H. (2017a). NHPP software reliability model considering the uncertainty of operating environments with imperfect debugging and testing coverage. *Applied Mathematical Modelling*, *51*, 68–85. doi:10.1016/j.apm.2017.06.034

Li, Q., & Pham, H. (2017b). A testing-coverage software reliability model considering fault removal efficiency and error generation. *PLoS One*, *12*(7). doi:10.1371/journal.pone.0181524 PMID:28750091

Li, X., Xie, M., & Ng, S. H. (2010). Sensitivity analysis of release time of software reliability models incorporating testing effort with multiple change-points. *Applied Mathematical Modelling*, *34*(11), 3560–3570. doi:10.1016/j.apm.2010.03.006

Liu, H.-W., Yang, X.-Z., Qu, F., & Shu, Y.-J. (2005). *A general NHPP software reliability growth model with fault removal efficiency*. Academic Press.

Luo, M., & Wu, S. (2018). A mean-variance optimisation approach to collectively pricing warranty policies. *International Journal of Production Economics*, *196*, 101–112. doi:10.1016/j.ijpe.2017.11.013

Malaiya, Y. K., Von Mayrhauser, A., & Srimani, P. K. (1993). An examination of fault exposure ratio. *IEEE Transactions on Software Engineering*, *19*(11), 1087–1094. doi:10.1109/32.256855

Mao, C., Xiao, L., Yu, X., & Chen, J. (2015). Adapting ant colony optimization to generate test data for software structural testing. *Swarm and Evolutionary Computation*, *20*, 23–36. doi:10.1016/j.swevo.2014.10.003

Mirjalili, S., Mirjalili, S. M., & Lewis, A. (2014). Grey wolf optimizer. *Advances in Engineering Software*, *69*, 46–61. doi:10.1016/j.advengsoft.2013.12.007

Mohanty, R., Ravi, V., & Patra, M. R. (2010). The application of intelligent and soft-computing techniques to software engineering problems: A review. *International Journal of Information and Decision Sciences*, *2*(3), 233–272. doi:10.1504/IJIDS.2010.033450

Musa, J. D. (1975). A theory of software reliability and its application. *IEEE Transactions on Software Engineering, SE-1*(3), 312–327. doi:10.1109/ TSE.1975.6312856

Musa, J. D. (1980). The measurement and management of software reliability. *Proceedings of the IEEE, 68*(9), 1131–1143. doi:10.1109/PROC.1980.11812

Musa, J. D. (1991). Rationale for fault exposure ratio K. *Software Engineering Notes, 16*(3), 79. doi:10.1145/127099.127121

Musa, J. D. (2004). *Software reliability engineering: more reliable software, faster and cheaper.* Tata McGraw-Hill Education.

Musa, J. D., Iannino, A., & Okumoto, K. (1987). *Software reliability: Measurement, prediction, application. 1987.* McGraw-Hill.

Naixin, L., & Malaiya, Y. K. (1996). Fault exposure ratio estimation and applications. *Proceedings of ISSRE'96: 7th International Symposium on Software Reliability Engineering.*

Nasar, M., Johri, P., & Chanda, U. (2013). A differential evolution approach for software testing effort allocation. *Journal of Industrial and Intelligent Information, 1*(2).

Pachauri, B., Dhar, J., & Kumar, A. (2015). Incorporating inflection S-shaped fault reduction factor to enhance software reliability growth. *Applied Mathematical Modelling, 39*(5-6), 1463–1469. doi:10.1016/j.apm.2014.08.006

Pai, P.-F. (2006). System reliability forecasting by support vector machines with genetic algorithms. *Mathematical and Computer Modelling, 43*(3-4), 262–274. doi:10.1016/j.mcm.2005.02.008

Pham, H. (1996). A software cost model with imperfect debugging, random life cycle and penalty cost. *International Journal of Systems Science, 27*(5), 455–463. doi:10.1080/00207729608929237

Pham, H. (2014). A new software reliability model with Vtub-shaped fault-detection rate and the uncertainty of operating environments. *Optimization, 63*(10), 1481–1490. doi:10.1080/02331934.2013.854787

Pham, H., Nordmann, L., & Zhang, Z. (1999). A general imperfect-software-debugging model with S-shaped fault-detection rate. *IEEE Transactions on Reliability, 48*(2), 169–175. doi:10.1109/24.784276

Pham, H., & Zhang, X. (1999). A software cost model with warranty and risk costs. *IEEE Transactions on Computers*, *48*(1), 71–75. doi:10.1109/12.743412

Rinsaka, K., & Dohi, T. (2005). Determining the optimal software warranty period under various operational circumstances. *International Journal of Quality & Reliability Management*, *22*(7), 715–730. doi:10.1108/02656710510610857

Roy, P., Mahapatra, G., & Dey, K. (2014). An NHPP software reliability growth model with imperfect debugging and error generation. *International Journal of Reliability Quality and Safety Engineering*, *21*(02), 1450008. doi:10.1142/S0218539314500089

Saed, A. A., & Kadir, W. M. W. (2011). *Applyisng particle swarm optimization to software performance prediction an introduction to the approach*. Paper presented at the 2011 Malaysian Conference in Software Engineering. 10.1109/MySEC.2011.6140670

Shanmugam, L., & Florence, L. (2013). *Enhancement and comparison of ant colony optimization for software reliability models*. Academic Press.

Sharma, D. K., Kumar, D., & Gautam, S. (2018). Flexible Software Reliability Growth Models under Imperfect Debugging and Error Generation using Learning Function. *Journal of Management Information and Decision Sciences*, *21*(1), 1–12.

Sheta, A., & Al-Salt, J. (2007). Parameter estimation of software reliability growth models by particle swarm optimization. *Management*, *7*, 14.

Song, X., Tang, L., Zhao, S., Zhang, X., Li, L., Huang, J., & Cai, W. (2015). Grey Wolf Optimizer for parameter estimation in surface waves. *Soil Dynamics and Earthquake Engineering*, *75*, 147–157. doi:10.1016/j.soildyn.2015.04.004

SourceForge.net. (2008). *An Open Source Software Website*. Author.

Srivastava, P. R., & Kim, T.-h. (2009). Application of genetic algorithm in software testing. *International Journal of Software Engineering and Its Applications*, *3*(4), 87–96.

Tyagi, K., & Sharma, A. (2014). An adaptive neuro fuzzy model for estimating the reliability of component-based software systems. *Applied Computing and Informatics*, *10*(1-2), 38-51.

Williams, D. P. (2007). Study of the warranty cost model for software reliability with an imperfect debugging phenomenon. *Turkish Journal of Electrical Engineering and Computer Sciences*, *15*(3), 369–381.

Windisch, A., Wappler, S., & Wegener, J. (2007). Applying particle swarm optimization to software testing. *Proceedings of the 9th annual conference on Genetic and evolutionary computation.* 10.1145/1276958.1277178

Yamada, S., Ohba, M., & Osaki, S. (1984). S-shaped software reliability growth models and their applications. *IEEE Transactions on Reliability, 33*(4), 289–292. doi:10.1109/TR.1984.5221826

Yamada, S., & Tamura, Y. (2016). *Software Reliability. In OSS Reliability Measurement and Assessment.* Springer. doi:10.1007/978-3-319-31818-9

Yamada, S., Tokuno, K., & Osaki, S. (1992). Imperfect debugging models with fault introduction rate for software reliability assessment. *International Journal of Systems Science, 23*(12), 2241–2252. doi:10.1080/00207729208949452

Zhang, X., Teng, X., & Pham, H. (2003). Considering fault removal efficiency in software reliability assessment. *IEEE Transactions on Systems, Man, and Cybernetics. Part A, Systems and Humans, 33*(1), 114–120. doi:10.1109/TSMCA.2003.812597

Chapter 2
Modified Multi–Grey Wolf Pack for Vital Sign–Based Disease Identification

Nabanita Banerjee
Techno Main Salt Lake, Kolkata, India

Sumitra Mukhopadhyay
Institute of Radio Physics and Electronics, University of Calcutta, India

ABSTRACT

Noninvasive process of vital sign identification and design of low-cost decision-making system for the betterment of rural health care support is a prime facet of research. The identification of bio-signals from different sensors, noise removal, signal processing, and decision making requires the use of sophisticated expert system. In this chapter, the authors propose a modified multi grey wolf pack optimization technique (MMGWO) for better generalization and diversification. The basic model has been modified using net energy gain of the individual wolf in the process of hunting. Here multiple packs of wolves are considered with simultaneous sharing and conflict among them. The performance of the proposed technique is tested on 23 well known classical benchmark functions, CEC 2014 benchmark problem set along with classical real-life applications. The experimental results and related analysis show that the proposed MMGWO is significantly superior to other existing techniques.

DOI: 10.4018/978-1-7998-1718-5.ch002

INTRODUCTION

Conventional diagnostic practices are very time consuming, labor-intensive and these require elaborate infrastructures and on-field expertise. It is difficult to set up such facilities in remote places and it is further difficult to obtain expert medical team for those places to cater the medical need for the general people with the available medical infrastructure. In such a scenario, design of non-contact and/or noninvasive medical support system may be considered as a primary target to connect those places for the minimum health care support. The implementation of such a scheme requires the widespread use of the sophisticated and automated expert system for disease detection. Here, basic bio signals are collected by different noninvasive methods using ensemble of sensors. Noises are removed from those signals; the signals are processed, decisions are taken and they are conveyed to medical experts present far away for better advice. The above methodology requires the design of an expert system with sophisticated sensor ensemble which collects various pathological parameters like blood pressure, pulse rate, SpO_2, perfusion rate, activity profile etc and they are transmitted to the signal processing module. The signals collected amidst real life environment needs robust and advanced level of signal processing algorithms. After getting the processed signals the hidden complex features are extracted and those non-redundant optimal set of features act as a fundamental backbone for first level of disease diagnosis and screening by decision making system. Now all the above modules like signal processing module, feature extraction module and above all the decision making module in individual stages require robust optimization algorithms for balanced and automated operation to cater the non-invasive remote medical functionality in remote area. The feature extraction module requires the application of robust optimization algorithm for feature extraction. Also the decision making system requires different mechanism like clustering, classification of data and parametric optimization for effective decision making and successively, a routine availability of minimum medical facilities may be ensured early in time to the people of remote places before the situation becomes a serious issue. Therefore, it may be observed that at all individual level of such kind of expert system design requires robust, auto adjustable, optimization algorithms with dedicated functionalities at different stages and the performances of those algorithms are finally going to be the driving factor in deciding the effectiveness of the support system. This will eventually decide the success of the system in providing the medical facility to the remote and rural healthcare paradigm. Inspired by the above requirement, in this book chapter, we propose to work on the most important aspect of such systems, i.e., on the design and development of sophisticated, simple but noble optimization algorithm and that may act as a back bone of any automated non-invasive health diagnostic system. Also, the above discussion may be supported in the review section where we find the immense use of optimization algorithms in the design and development of non-invasive technology-based diagnostic methods.

Optimization means the process of finding the best solution(s) of a particular problem(s). Due to simple and flexible nature and derivation free mechanism, the metaheuristic optimization process has gained immense popularity over the past few decades. In mathematical optimization, a metaheuristic is a higher-level procedure or heuristic designed to find a good solution to an optimization problem, especially with incomplete or imperfect information or limited computational capacity. Metaheuristics may make few assumptions about the optimization problem being solved. They are used in several domains such as industry, engineering, business, computer sciences, etc. Meta-heuristic algorithms are basically population based. All the algorithms have been inspired by the physical phenomena, animals' behaviors, or evolutionary concepts, etc. A variety of metaheuristics, such as genetic algorithm (GA) (Bonabeau, Dorigo, & Theraulaz, G.,1999) ant colony optimization (ACO) (Dorigo, Birattari, & Stutzle, 2006) particle swarm optimization(PSO)(Kennedy, & Eberhart 1995), firefly algorithm (Yang,2010) artificial bee colony (ABC) optimization(Karaboga & Basturk, 2007), spider monkey optimization (SMO) (Bansal, Sharma, Singh & Maurice, 2014), whale optimization (Mirjalili & Lewis, 2016) have been proposed and successfully applied in many engineering fields. These algorithms have great potential to deal with real world optimization problems. The first step of population-based optimization is to make a set of random initializations. Each solution is considered as a candidate solution of the assigned problem. It utilizes the concept of objective functions and objective values. The algorithm updates the candidate solution according to the best solution value to improve them. The solutions are again assessed by objective values and their relevant fitness values are redefined. This process is continued until the satisfactory result obtained. Lastly, the best solution is noted as the best approximation of global optimum.

There are three types of population-based algorithms. The first main branch of meta-heuristics or population-based algorithm is an evolutionary based technique. The second branch of meta-heuristics or population-based algorithm is physics-based technique. The third meta-heuristics or population-based algorithm is swarm intelligence (SI) based technique. The most well-known SI technique is Particle Swarm Optimization (PSO), proposed by Kennedy and Eberhart. It is based on the natural behavior of bird flocking. It utilizes the concept that a number of particles (candidate solution) which fly around the search space to find the best solution. ACO is also a popular technique in this branch. It is proposed by Dorigo et al.in2006 and is based on the social behavior of the ant in an ant colony. The basic aim of this algorithm is to find the shortest path between the colony of ants and the food source. Another popular meta-heuristic algorithm is Bat Algorithm (BA) proposed by Xin She Yang (Yang, 2011). Bat uses the echolocation for hunting the prey. Grey wolf optimization algorithm (Mirjalili, Mirjalili & Lewis, 2014) is the other most popular SI algorithm which is invented in 2014.The other some of the popular SI techniques are:

1. Marriage in Honey Bees Optimization Algorithm (MBO) (Celik & Ulker, 2013)
2. Artificial Fish-Swarm Algorithm (AFSA) (Azizi, 2014)
3. Glowworm Swarm Optimization (GSO) (Liao, Kao & ShanLi, 2011)
4. Cuckoo Search Algorithm (CSA) (Basu & Chowdhury, 2013)
5. Artificial Bee Colony (ABC) (Karaboga & Basturk, 2007)
6. Spider Monkey Optimization (SMO) (Bansal, Sharma,Singh & Maurice)
7. Whale Optimization (WO) (Mirjalili & Lewis, 2016)
8. Ant Lion Optimizer (ALO) (Mirjalili, 2015)
9. Fruit Fly Optimization algorithm (FOA) (Wang, Zheng & Wang, 2013) etc.

Bio-inspired optimization techniques have become very popular because of their applicability to a wide range of optimization problems. All the of this category are inspired by various physical phenomena like food searching, hunting, bird flocking, animals' behaviors, or the evolutionary process, etc. Immature convergence and stagnation at local optima (LO) are some common disadvantages of bio-inspired optimization techniques. To overcome these drawbacks, researchers are continually in the process of exploring new algorithms with optimal set of features. One of the popular and comparatively new bio-inspired optimization techniques is Grey Wolf Optimization proposed by Mirjaliliet.al. in 2014 to overcome the drawbacks of the existing algorithms. GWO have several advantages like: it is easily adapted to several optimization problems. No acquired information is needed in the initial search. A few parameters of GWO are initialized, which is superior in avoiding local optima. GWO algorithm is simple, flexible, adaptable, usable, and stable. It also has some disadvantages like low solving precision, slow convergence, and bad local searching ability. Therefore, studying both the advantages and disadvantages of GWO, we choose to modify the basic algorithm, increase the optimal balance of exploration and exploitation and propose a modified algorithm based on GWO. The GWO algorithm is inspired from the basic hunting procedure and social hierarchy of the grey wolf in nature. It shows immense potential in the domain of search problem and global optimization. Grey wolves live in a pack of approximately five to eleven wolves in the forest. They are considered as apex predators that mean they are at the top of the food chain. Social hierarchy is the main feature of their pack. For hunting the prey and to maintain the discipline within the pack, they categorize their group into four types of wolves. The alpha is considered as a leader of the total pack. The second level of wolves is Beta wolves. Omega is the lowest ranking grey wolves. If the wolf does not belong to any of the above categories is called delta. In grey wolf hunting strategy-based algorithms, alpha is the considered the best solution, beta is the second best and delta is the third best solution. In this algorithm, the first three solutions are archived and the other search agents (omegas) update their positions according to the positions of the best search agents. The grey wolves follow the three steps. These are –

1. Tracking, chasing and approaching the prey.
2. Pursuing, encircling and harassing the prey.
3. Attacking the prey.

However, there are two crucial issues present in working of GWO that need attention. GWO works around the global best solution, which makes it vulnerable to getting stuck in local optima. Another issue is that GWO loses its convergence speed as the algorithm progresses. GWO has a chronic disadvantage in loosing diversity due to its high dependency on the three best (alpha, beta, delta wolves) solutions found in the accumulative search. For these reasons GWO is greedy in nature and it does not share knowledge from other solutions. Now, for better generalization, convergence and enhancement of fundamental characteristics, the hunting procedure of GWO is modified. To introduce and maintain an optimum balance between exploration and exploitation in the search space, Modified Multi Grey Wolf Packabbreviated as MMGWO is proposed in this chapter. The aims of the proposed MMGWO are better generalization, search procedure and diversification. The main features of the proposed algorithm are listed as follows:

1. The multipack GWO has been introduced instead of single pack for diversification.
2. In a single pack, the wolves show hierarchy as alpha, beta, gamma and delta wolves. In the proposed modification instead of considering single wolf as alpha, beta and delta wolf each, top a_p wolves of pth pack are selected as alpha wolves and they are termed as the member of alpha pack. Next top b_p wolves are selected as members of beta pack and next top d_p wolves are selected as the member of delta packs of pth pack. Rests of the wolves in the population are omega wolves. Here P numbers of packs are present in the search space and p varies from 1 to P. During updation, one of the wolves is randomly selected, each from alpha, beta and delta pack.
3. In the process of hunting, the grey wolf spent energy for searching their prey. Usually in a forest there are multiple preys present at a time. The larger prey being a source of high energy is more preferable. But unfortunately, larger

prey might be few or at a greater distance, as compared to smaller prey. Hence, selection of prey for wolves involves a trade-off between the potential energy gain from the prey, and energy spent in catching it. Therefore, wolves are more likely to select prey that has the maximum potential to provide with the highest net energy intake. These foraging characteristics of wolves have been modeled and incorporated into the update equation to introduce a modification into the basic optimization process. This concept adds on more diversity in the optimal solution search. The convergence graphs for different functions are shown in Figure7, Figure 8, Figure 9, Figure 15 and Figure 16 and they also support the introduction of the proposed feature. Also, the basic structure of a single wolf pack is remodeled with overall energy gain within the pack.

4. In a forest multiple wolves pack may roam around and hunt for prey. The packs sometime share information among each other in terms of both mutual conflict and information sharing. In the proposed modification the multiple packs of GWO have been considered. Their mutual information sharing and conflict is modeled. The populations of the updated packs are crossed and new set of offspring (child) is generated. The new generation compares their position with the position of two parental populations. The best solution is absorbed in the next iteration.

This process continues until the best optimal results are obtained. To investigate the performance of the proposed method, MMGWO is tested on 23 classical benchmarks, CEC 2014 benchmark problems and some real time applications. It gives promising results with respect to other existing conventional algorithms. The convergence graphs (Figure 7, Figure 8, Figure9, Figure 15 and Figure 16) and the bar charts (Figure 4, Figure 5, Figure 6, Figure 10, Figure 11, Figure 12, Figure 13 and Figure 14) support the enhancement of the basic GWO.

The rest of the chapter is demonstrated as follows: Background section explains the previous work done in this field. The conventional GWO techniques are illustrated in the next section. After this, the proposed MMGWO is described in detail. In result section, the MMGWO along with other optimizers have been applied to the classical benchmark problems and CEC problem set. Then, the next part of chapter demonstrates the use of MMGWO in real time problem solving. Finally, the chapter ends with the conclusion and future work.

BACKGROUND

In recent two decades, many SI optimization techniques have been introduced to solve constrained optimization, unconstrained optimization, and engineering design problems due to their simple and flexible nature and derivation free mechanism. These techniques have been inspired by the physical phenomena, animals' behaviors, or evolutionary concepts, etc and they use the concept of group intelligence along with individual intelligence. The popular techniques in this category are particle swarm optimization (PSO), firefly algorithm, artificial bee colony (ABC), spider monkey optimization (SMO), whale optimization etc. Recently many researches with different variations of the basic algorithms have proposed in this domain. For example, in 2014, Amir H. Gandomi, Xin-She Yang (Amir, Gandomi & Yang, 2014) developed chaotic bat algorithm which improves its global search mobility for robust global optimization. In this paper chaos into the bat algorithm (BA) is introduced for increasing global search. In 2017, Hema Banati, Reshu Chaudhary (Banati & Chaudhary, 2017) proposed multimodal variant of the bat algorithm with improved search (MMBAIS). This multi-modal characteristic is obtained by introducing the foraging behavior of bats. In this paper an additional parameter has been proposed to refine the search capabilities. Y. Chen et.al. (Chen, 2017) developed a new approach of PSO with two differential mutation. In this paper, authors proposed a velocity update scheme with learning model. Two differential mutation operators with different characteristics are applied here. Furthermore, the authors adopted a new structure with two layers and two swarms. For better exploitation in this paper, a dynamic adjustment scheme for the number of particles in two sub-swarms is proposed. This scheme is simple and effective. N. Lynn and P. Nagaratnam Suganthan (Lynn & Suganthan, 2017) presents a new concept of PSO entitled as -ensemble particle swarm optimizer. In this algorithm a self-adaptive scheme is proposed to identify the top algorithms by learning from their previous experiences.

Grey wolf optimization (GWO) (S. Mirjalili et al., 2014) is another population-based algorithm which utilizes the hunting mechanism of grey wolves. Now a day, this algorithm has gained immense popularity. Recently, GWO algorithm has been used in different modes. In order to increase the efficiency of GWO, levy flight (LF) and greedy selection strategies are added by A. A. Heidari and P. Pahlavani for solving both global and real-world optimization problems (Heidari & Pahlavani, 2017). Another concept is developed by Vijay Kumar and Dinesh Kumar (Kumar & Kumar, 2017) for preventing premature exploration and convergence of the GWO algorithm on optimization problems. They added prey weight and astrophysics-based concepts together with their algorithm.

In 2016, Medjahed et. al. (Medjahed, 2016) developed a GWO-based procedure for the hyper spectral band selection task. In 2018, Chao Lua et. al. proposed a concept of GWO entitled as grey wolf optimizer with cellular topological structure (Chao, Liang, JinYi,2018). In another work (Jayabarathi, Raghunathan, Adarsh, Suganthan, 2016) conventional GWO has been hybridized with mutation and crossover mechanisms to handle the economic dispatch problems. In 2018, M.A. Al-Betar et.al. proposed a new GWO algorithm which is based on the natural selection methods for the grey wolf optimizer.

In 2018 R.A Ibrahim et.al. proposed chaotic opposition-based grey-wolf optimization algorithm which is based on differential evolution and disruption operator for improving the exploration ability of the GWO.In this paper, the authors introduced the chaotic logistic map, the opposition-based learning (OBL),the differential evolution (DE), and the disruption operator (DO)for enhancement of the exploration ability of the basic GWO. Sen Zhanga and Yongquan Zhou (2017) proposed a new technique for matching the template using the GWO. Here the authors proposed a hybrid method of GWO. In this paper GWO and lateral inhibition (LI) are hybridized for solving complex template matching problems. In 2016, M.R. Shakarami et.al. developed a method for designing wide-area power system stabilizer (WAPSS) utilizing GWO algorithm (Shakarami et.al.,2016). Mingjing Wang and Huiling Chen (Wang & Chen, 2017) proposed a new kernel extreme learning machine (KELM) parameter tuning strategy based on GWO algorithm. Here GWO algorithm is used to construct an effective KELM model for bankruptcy prediction.

In 2018, L.K. Panwaret.al. (Panwar et.al., 2018) developed binary GWO for large scale unit commitment problem. E. Emary .et. al. (2016) gave an idea of feature selection using binary GWO. In this paper, a binary GWO is develop and utilized to select an optimal feature subset for classification purposes. Lu, Chao (2017) developed a hybrid multi-objective grey wolf optimizer for dynamic scheduling in a real-world welding industry. S. Mirjalili et.al. (Mirjalili et.al., 2016) developed a multi-objective grey wolf optimization in 2016.

A. N Jadhav and N. Gomathi (2018) proposed a hybrid model of the GWO. Here GWO is hybridized with whale optimization for data clustering. This algorithm uses the minimum fitness measure and the fitness measure depends on three constraints. They are inter-cluster distance, intra-cluster distance, and the cluster density.

Our overall target of the work is to provide medical-resource under-served rural sector and development of portable, affordable medical devices involving biosensor, low-power miniature multi-parameter physiological signal acquisition hardware, real-time low-complexity bio signal processing techniques for assessing signal quality and extracting vital signs. The extracted vital signs may be used for point-of-care (POC) detection of different diseases like heart disease; lung related diseases, cancer, stress related diseases etc.

In 2017, Anita Sahoo and Satish Chandra (Sahoo, A., & Chandra, S. 2017) developed a multi-objective GWO for improved cervix related disease detection.

Recent years cervical cancer is one of the most important and frequent cancers. But it can be cured if diagnosed in the early stage. In this paper a novel effort is shown towards effective characterization of cervix lesions.

In 2018, S. Lahmiri, A.Shmuel (Lahmiri & Shmuel, 2019) developed a machine learning process for diagnosing Alzheimer's disease (AD). The aim of this work is to evaluate the features which help us to classify AD patients and healthy control subjects using several machine learning classifiers. Q. Li et.al. (2017) developed an enhanced GWO for feature selection wrapped kernel extreme learning machine for medical diagnosis. Here, a predictive framework is developed by integrating an improved GWO (IGWO) and kernel extreme learning machine (KELM). The proposed IGWO feature selection mechanism is utilized for finding the optimal feature subset of medical data. In this approach, genetic algorithm (GA) was firstly introduced to evaluate the diversified initial positions, and then GWO was introduced to update the current positions of the population.

M. Kachuee, M. M. Kiani, H. Mohammadzade, M. Shabany (2015) developed a method for non-invasive cuff-less estimation of systolic and diastolic blood pressures. The work has been accomplished by extraction of several physiological parameters from photoplethysmography (PPG) signal. Then it utilizes signal processing and machine learning algorithms. In 2015, Selim Dilmac and Mehmet Korurek proposed modified artificial bee colony (MABC) for classification of the electrocardiogram (ECG) heartbeat.

The psychological stress is an important health problem in today's world and the state of stress acts as a major catalyst to a number of disease processes like cardiovascular disease, depression, cancer etc. and it leads to a social catastrophe in many households with low or moderate source of income. According to World Health Organization (WHO) report, depression will be ranked second in the ten leading causes of the global burden of disease by 2020.Therefore, non-invasive early detection of stress is becoming an important part of research with vital signs.

In 2014, Nandita Sharma and Tom Gedeon modeled a stress signal. The aim of this paper is to estimate an objective stress signal for an observer of a real-world environment stimulated by meditation. A computational stress signal predictor system is made which was developed based on a support vector machine, genetic algorithm and an artificial neural network to predict the stress signal from a real-world data set.

In 2018, Amandeep Cheema and Mandeep Singh (Cheema & Singh, 2019) proposed an application of phonocardiography (PCG) signals for psychological stress detection. The aim of this research paper is to design a new framework using PCG signal to detect psychological stress based on non-linear entropy-based features extracted. In this paper the authors used heart sound or phonocardiography (PCG) to detect psychological stress instead of ECG signals. Temporal correlation exists between R-peak of ECG signal and S1 peak of PCG signal. So here the authors used the PPG signal to detect psychological stress rather than using ECG based HRV signals. Here EMD approach has been used to assess on-linear entropy-based features which are captured by non-linear dynamics of cardiac activities from PCG signals. The authors made a framework of portable device with an electronic microphone. It can serve as a homecare-based psychological stress measurement system for timely diagnosis. In future, they can develop an expert system using soft computing methods to detect and analysis of cardiac activities from heart sound as a pre-screening method where the medical expert or sophisticated equipment are absent. Table1 shows the different applications and drawbacks of the methods which used different optimization algorithms and machine learning techniques.

GREY WOLF OPTIMIZATION: BASIC

Grey wolf optimization algorithm is a population based meta-heuristic algorithm which imitates the hunting procedure of the pack of grey wolves. They belong to the elite family of the food chain and thus maintain a social hierarchy as shown in Figure1. They like to live in a pack and the group size is 5-12 on average. The alpha is considered as a leader of the total pack. The second levels of wolves are Beta. Omega is the worst grey wolves.

The leaders may be a male or a female, called alpha. The alpha makes decisions about hunting, sleeping place, time to wake, and so on. An alpha is followed by the other wolves in the pack. When gathering shield, the entire pack acknowledges the alpha by holding their tails down. The orders should be followed by the pack (Mech LD., 1999).

The second level grey wolves are beta. The betas help the alpha in decision-making or other pack activities. Beta is probably the best candidate to be the alpha in case one of the alpha wolves passes away or becomes very old. The beta wolf should respect the alpha, but commands the other lower-level wolves as well. The lowest ranking grey wolf is omega. The omega plays the role of scapegoat. If a wolf is not an alpha, beta, or omega, he/she is known as delta. Delta wolves are lower ranking wolves than alphas and betas, but they dominate the omegas. In addition to the social hierarchy, group hunting is another interesting social behavior of grey wolves. The main characteristics of grey wolf hunting (Mirjalili, 2014) are shown in Figure 2 and they are described as follows:

1. Chasing, tracking, and approaching the prey.
2. Pursuing, encircling, and harassing the prey until it stops moving.
3. Attacking towards the prey.

Table 1. Medical applications-oriented algorithms and their drawbacks

Sl.No	Name of the Paper	Applications of the Paper	Comments
1	"Multi-objective grey wolf optimizer for improved cervix lesion classification.(Sahoo, Anita & Chandra, 2017)	Reflects a novel effort towards effective characterization of cervix lesions.	Here only five features are extracted. The automated feature selection may result in achieving better accuracy. Two multi objective GWO algorithms are used. It makes the algorithm more involved to implement.
2.	Performance of machine learning methods applied to structural MRI and ADAS cognitive scores in diagnosing Alzheimer's disease.(Lahmiri, Salim & Shmuel, 2018)	Classify AD patients and healthy control subjects using several machine learning classifiers.	The data collections are rigorous job. The combination of cortical models and cortical metrics gives lower performance. This can be allocated to features redundancy.
3.	An enhanced grey wolf optimization-based feature selection wrapped kernel extreme learning machine for medical diagnosis. (Qiang, et al.,2017)	Utilized for finding the optimal feature subset of medical data.	Only binary version of GWO is used to combine with GA. The kernel extreme learning machine (KELM) classifier is introduced to evaluate the fitness value. The hybrid structure makes the model more complicated.
4.	Cuff-less high-accuracy calibration-free blood pressure estimation using pulse transit time. (Kachuee & Mohamad, et al., 2015)	Developed a method for non-invasive cuff-less estimation of systolic and diastolic blood pressures. This paper is accomplished by extraction of several physiological parameters from PPG signal.	In this paper, some assumptions are taken for blood pressure (BP) measurement.
5.	ECG heart beat classification method based on modified ABC algorithm. (Dilmac & Korurek, 2015)	Classify of the ECG Heartbeat.	Balanced data set is needed for implementation of this algorithm for ECG heartbeat classification. But it is impossible for finding the balanced dataset. Different data sets, unbalanced sample numbers in different classes have effects on classification result.
6.	Modeling a stress signal. (Sharma & Gedeon, 2014)	Estimation an objective stress signal for an observer of a real-world environment stimulated by meditation.	Data collections are very difficult. The signals are collected from a patient when he or she is in a meditative state. In this paper GA and SVM hybrid model is used.
7.	An application of phonocardiography signals for psychological stress detection using non-linear entropy-based features in empirical mode decomposition domain. (Cheema & Singh,2018)	Using heart sound or phonocardiography (PCG) to detect psychological stress instead of ECG signals.	In real time easy measurement of phonocardiography signals is difficult. Physical stress affects the Phonocardiography signals Extraction of optimal feature set from this signal in real time involves a set of procedures.

In the conventional GWO, the encircling procedure is modeled as follows (Mirjalili, 2014).

$$\vec{D} = \vec{C} . \vec{X}_{pr} (t) - \vec{X} (t) \tag{1}$$

$$\vec{X}(t+1) = \vec{X}_{pr}(t) - \vec{A} . \vec{D} \tag{2}$$

The vectors $\vec{A}$ and $\vec{C}$ are calculated as follows;

$$\vec{A} = 2.\vec{a}.\vec{r}_1 - \vec{a} \tag{3}$$

$$\vec{C} = 2.\vec{r}_2 \tag{4}$$

Where t is the number of iterations, $\vec{A}$ and $\vec{C}$ are random vectors, $\vec{X}_{pr}$ is the location of the prey, and $\vec{X}$ is the position of wolves D indicates the distance between wolves and prey. In grey wolf strategy, alpha is the best solution and beta and delta are the second and the third best solutions, respectively. In this algorithm, the first three solutions are archived and the rest of the wolves (omegas) update their positions according to the positions of the best solutions. The algorithm of GWO is shown in Algorithm 1.

The hunting procedure of the wolf is modeled as follows.

$$\vec{D}_\alpha = \vec{C}_1 . \vec{X}\alpha - \vec{X}, \vec{D}_\beta = \vec{C}_2 . \vec{X}_\beta - \vec{X}, \vec{D}_\gamma = \vec{C}_3 . \vec{X}_\delta - \vec{X} \tag{5}$$

Figure 1. Hierarchy of grey wolf

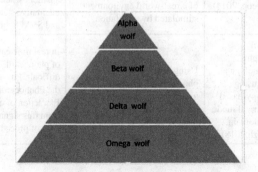

$$\vec{X}_1 = \vec{X}_\alpha - \vec{A}_1 . \vec{D}\alpha, \ \vec{X}_2 = \vec{X}_\beta - \vec{A}_2 . \vec{D}_\beta, \ \vec{X}_3 = \overrightarrow{X_\delta} - \vec{A}_3 . \vec{D}\gamma \tag{6}$$

$$\vec{X}(t+1) = \frac{\vec{X}_1 + \overrightarrow{X_2} + \overrightarrow{X_3}}{3} \tag{7}$$

where $\vec{X}\alpha$; $\vec{X}_\beta$; $\overrightarrow{X_\delta}$ represent the locations of alpha, beta and delta wolves respectively; C_1; C_2; C_3, A_1, A_2 ; A_3 denote random vectors, and $\vec{X}$ signifies the location of the present solution.

MODIFIED MULTI GREY WOLF PACK GWO (MMGWO)

Grey wolf optimization (GWO) has several advantages like:

- The algorithm may be easily implemented.
- Space requirement of the algorithm is less.
- Convergence is faster because of the continuous reduction of the search space. The decision variables are less and only two parameters A and C are to be adjusted mainly.

Figure 2. Hunting behavior of grey wolves (A) chasing, approaching, and tracking prey (B) pursuing (C) harassing (D) encircling (E) stationary situation and attack

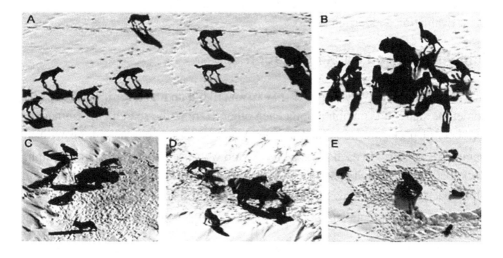

However, it has also some disadvantages like low solving precision, slow convergence, and bad local searching ability. Therefore, studying both the advantages and disadvantages of GWO, we choose to modify the basic algorithm, increase the optimal balance of exploration and exploitation and propose a modified algorithm entitled as Modified Multi Grey Wolf Pack (MMGWO) algorithm. The aims of the proposed algorithm are better generalization, search procedure and diversification.

In this work, first we have tried to introduce the concept of multipack GWO instead of single pack for diversification. Secondly, group hierarchy of the GWO has been remodeled. In a single pack, wolves are generally put in a hierarchical order like alpha, beta, delta and omega wolves where top three wolves are named as alpha, beta and delta wolves respectively and rest of the wolves are in omega category. Here, in the proposed modification instead of considering single wolf as alpha, beta or delta wolf, top a_p wolves of pth pack are selected as alpha wolves and they are designated with a special status like the member of alpha pack. Next top b_p wolves are selected as members of beta pack and next top d_p wolves are selected as the member of delta packs of pth pack. Rests of the wolves in the population are omega wolves of pth pack. Here P numbers of packs are present in the search space and p varies from 1 to P. The updation of the wolf pack members are done based on the random members selected from alpha pack, beta pack and delta pack respectively instead of alpha, beta and delta wolves. Lastly, the updation of individual wolf is remodeled by introducing the overall energy gain concept of individual wolf. The development of MMGWO algorithm is described in figure 3 and algorithm 2.

The total work is described by the following steps:

1. Initialization and pack formation
2. Energy based updation of the population
3. Information exchange between the wolf packs
4. Evaluation and termination

Initialization and Pack Formation

In a forest, multiple wolves' packs may roam around and hunt for prey. The packs sometime share information among each other. Here in the search environment, P numbers of packs are initialized having M_p wolves in each pack along with its parametric settings. In each pack, again the top wolves are distributed among alpha pack, beta pack and delta pack respectively. The rest of the wolves are in omega. In alpha pack, top a_p wolves of pth pack are selected. Similarly, b_p and d_p are the number of next top beta and delta pack members of pth pack respectively. Here P numbers of packs are present in the search space and o_p is the number of members who are omega wolves in the pth pack.

In our implementation section, we have considered two pack implementation approaches, i.e., P=2.

$$M_p = a_p + b_p + d_p + o_p, \ p = 1,\ldots\ldots,P \tag{8}$$

The location of the ith pack member of pack p, $\vec{X}_i^p(t)$ is initialized within the upper bound ub and lower bound lb as follows,

$$\vec{X}_i^p(t) = (ub - lb).* rand(1, Dim) + lb \tag{9}$$

p= 1, 2,……,P (Number of pack);

$i = 1, 2, ..., M_p$ (Members of the pth pack) and rand () function generates Dim-dimensional uniformly distributed random vector.

Energy Based Updation of the Population

The wolves in each pack concurrently and independently explore the entire search space for optimal results. The wolves update its hunter towards the prey as described in the previous section using equation (1) to (7). In this paper a modification is done on the selection scheme of alpha, beta and delta wolf. Instead of considering the best, second best and third best wolf as alpha, beta and delta wolf, respectively, randomly one wolf is selected each from alpha pack, beta pack and delta pack. This concept adds on more diversification in the search space and is modeled as follows.

$$\alpha pack^p = randinit(1, a_p); \tag{10}$$

$$\beta pack^p = randinit(a_p, a_p + b_p); \tag{11}$$

$$\delta pack^p = randinit(a_p + b_p, a_p + b_p + d_p); \tag{12}$$

where $\alpha pack^p, \beta pack^p$, and $\delta pack^p$ are the randomly selected members from the alpha pack, beta pack and delta pack respectively for further updation of pth pack.

The hunting procedure of the wolf is modeled as follows.

$$\vec{D}^p_{\alpha pack} = \vec{C}_1 . \vec{X}^{\alpha pack^p} - \vec{X}^p, \vec{D}^p_{\beta pack} = \vec{C}_1 . \vec{X}^{\beta pack^p} - \vec{X}^p, \vec{D}^p_{\delta pack} = \vec{C}_1 . \vec{X}^{\delta pack^p} - \vec{X}^p \tag{13}$$

$$\vec{X}^p_1 = \vec{X}^{\alpha pack^p} - \vec{A}_1 . \vec{D}^p_{\alpha pack}, \ \vec{X}^p_2 = \vec{X}^{\beta pack^p} - \vec{A}_1 . \vec{D}^p_{\alpha pack}, \ \vec{X}^p_3 = \vec{X}^{\delta pack^p} - \vec{A}_1 . \vec{D}^p_{\delta pack}, \tag{14}$$

$$\vec{X}^p(t+1) = \frac{\vec{X}^p_1 + \vec{X}^p_2 + \vec{X}^p_3}{3}, \ \text{p=1, 2, ..., P} \tag{15}$$

where $\vec{X}^{\alpha pack^p}$, $\vec{X}^{\beta pack^p}$ and $\vec{X}^{\delta pack^p}$ represent the locations of selected wolves from alpha, beta and delta packs respectively; C_1; C_2; C_3, A_1, A_2 ; A_3 denote random vectors, and $\vec{X}^p(t)$ signifies the location of the present solution.

Again, this basic scheme sometime suffers from pre-mature convergence. To overcome these drawbacks, the basic model of a single wolf pack is also remodeled with the overall energy gain within the pack. This concept introduces more diversity in the optimal solution search. During hunting, the grey wolf first scans its environment for all possible preys and spent energy for searching their prey. Usually there are multiple preys present at a time. The larger prey being a source of high energy is more preferable. But unfortunately, larger prey might be few or at a greater distance, as compared to smaller prey. Hence, selection of prey for wolves involves a trade-off between the potential energy gain from the prey, and energy spent in catching it. Therefore, wolves are more likely to select prey that has the maximum potential to provide with the highest net energy intake. These foraging characteristics of wolves have been modeled and incorporated into the update equation to introduce a modification into the basic optimization process. This concept adds on more diversity in the optimal solution search. The net energy gain for i^{th} wolf in p^{th} pack is calculated as:

$$eG^p_{i,j} = \left| E^p_i - E^p_j \right| \tag{16}$$

$$r_{ij}^p = \vec{X}_i^p(t) - \vec{X}_j^p(t) = \sqrt{\left[\sum_{d=1}^{D}\left(x_{i,d}^p - x_{j,d}^p\right)^2\right]} \tag{17}$$

$$eS_{i,j}^p = r_{ij}^p \delta^p \tag{18}$$

$$\psi_{ij}^p = eG_{i,j}^p - eS_{i,j}^p + \frac{1}{2}m_i^p.v_i^p \tag{19}$$

Where E_i^p and E_j^p are the objective function values of ith and j[th] wolves of the pth pack and their difference, $eG_{i,j}^p$ gives possible energy gain. $eS_{i,j}^p$ gives the possible amount of energy spent in covering Cartesian distance (r_{ij}^p) between i[th] and j[th] wolves of the pth pack. Here δ^p is the energy consumption coefficient of the pth pack. ψ_{ij}^p gives net energy gain. m_i^p is the mass of the wolf and $.v_i^p$ is the velocity of the wolf in the pth pack and $\frac{1}{2}m_i^p.v_i^p$ represents the kinetic energy of the ith wolf.

The overall updated equation is given as follows.

$$\vec{X}^p(t+1) = \left(\frac{\vec{X}_1^p + \vec{X}_2^p + \vec{X}_3^p}{3}\right).\psi_{ij} \text{ when } lb \leq \vec{X}^p(t) \leq ub \tag{20}$$

Information Exchange Between the Wolf Packs

Once the exploration process is over, the evolved search agents from the packs are probabilistically crossed, new set of offspring are generated, and they are separately compared with the parents. If the generated offspring set is found to be better than the existing members of respective pack then the members of the pack are updated. In terms of the above procedure, exchange of information is accomplished within the packs which further enhance the diversification of the algorithm and the solution space is explored in greater extent.

Crossover $(\vec{X}_i^k, \vec{X}_j^m)$ if rand () <Pr $\tag{21}$

Where Pr is the probability factor and crossover () function implements the crossover between i[th] member of kth pack and j[th] member of mth pack

Evaluation and Termination

The process is continued until the stopping criterion like number of iteration or error is reached within desired threshold value. At the end of each iteration, elite solution $\vec{X}_{best}$ is archived for achieving optimal solution in the search process.

$$\left[fbest, bestpack \right] = \min \left\{ f\left(\vec{X}^{\alpha pack^1} \right), \left(\vec{X}^{\alpha pack^2} \right), ..., \left(\vec{X}^{\alpha pack^P} \right) \right\} \tag{22}$$

$$\vec{X}_{best} = \vec{X}^{\alpha pack^{bestpack}} \tag{23}$$

Where fbest is the fitness value of the best wolf among all the packs and best pack is the corresponding pack number.

Computational Complexity of MMGWO

In this section the computational complexity of the proposed algorithm is compared with some other established algorithms to ensure the effectiveness of the proposed algorithm. The order of complexity is given in terms of O notation on the basis of the process initialization; evaluation and update in Table 2 where *N, D, cost* and *iter_max* represent the population size, dimension of the problem, corresponding cost function and the maximum number of iterations respectively. The overall complexity of MMGWO is compared with PSO, DE, ABC and GWO algorithms (Agarwalla, P., & Mukhopadhyay, S. (2020)). The complexity of initialization and evaluate phase is O(ND*iter_max) and O(cost*N*iter_max) respectively. Apart from initialization and evaluation phase, update phase of MMGWO is divided into two subpart like- 1) the population is updated in terms of energy equations and 2) information sharing occurs via probabilistic crossover. So, the complexity of energy equation based update and probabilistic crossover section becomes O(ND*iter_max) and O (2*D*iter_max) respectively and thus the complexity of the overall update section becomes O(ND*iter_max). The overall complexity of the proposed MMGWO algorithm is O(FE*D+FE*cost). It shows that the order of complexity does not increase in the proposed method in comparison to other widely known algorithms.

Algorithm 1. Grey Wolf Optimization(GWO) algorithm

```
Initialization the grey wolf populations X (i=1,2, …n)
Initialize a, A and C
Calculate the fitness of each search agent
X =the best search agent
X =the second best search agent
X =the third best search agent
while (t<maximum number of iterations)
for each search agent
    update the position of the current search agent by
equation(1 to 7)
end for
update a,A and C
    calculate the fitness of all search agents
    update X, X, X
    t=t+1
end while
return X
```

EXPERIMENTAL SETUP

In this section, we have performed three sets of experiments to evaluate the performance of the proposed MMGWO. In the first set, the algorithm is examined on 23 classical benchmark (Mirjalili, 2014) problems. Out of which functions F01-F07 are classified as unimodal test cases and the rest are multimodal test cases. Here Dim indicates the dimension of the function, Range is the boundary of the function's search space, and f_{min} is the optimum value. We run the MMGWO algorithm 30 times for each function of classical problems. The statistical results, e.g. average value (Ave) and standard deviation (STD) of 30 runs each having 500 iterations for MMGWO algorithm and other methods are shown in Table 7.

The second set of experiment is done on CEC 2014 problem set which are listed in Table 6. The algorithm is tested 30 times for each function. Here the dimension of the evaluated function is 50 and functions evaluation (FE) is taken as 500000. The statistical results, e.g. average value (Ave) and standard deviation (STD) of 30 runs for MMGWO algorithm and other algorithms are shown in Table 8.

The third set of experiment is done on two classical engineering problem like pressure vessel design problem and wielded beam design problem. The proposed

Algorithm 2. Modified Multi Grey Wolf Pack GWO

```
%INITIALIZATION
```
Initialization the multi grey wolf pack populations $\vec{X}_i^p(t)$

($i=1,2,\dots M_p, p = 1,2,\dots\dots,P$)

Initialize a, A, C, M_p, P, D, m_i^p, lb, ub, a_p, b_p, d_p

```
%FITTNESS EVALUATION AND PACK FORMATION
```

Calculate the fitness of each search agent $\vec{X}_i^p(t)$ in each pack p

```
%IDENTIFY THE ALPHA PACK, BETA PACK, AND DELTA PACK MEMBERS IN
EACH PACK
```

$$\alpha pack^p = randinit\left(1, a_p\right);$$

$$\beta pack^p = randinit\left(a_p, a_p + b_p\right);$$

$$\delta pack^p = randinit\left(a_p + b_p, a_p + b_p + d_p\right);$$

$$\vec{X}^{\alpha pack^p}(t) = \vec{X}^p_{\alpha pack^p}(t); \qquad \vec{X}^{\beta pack^p}(t) = \vec{X}^p_{\beta pack^p}(t); \qquad \vec{X}^{\delta pack^p}(t) = \vec{X}^p_{\delta pack^p}(t)$$

```
%START OF LOOP
while (t<maximum number of iterations)
for p=1 to P do
%ENERGY BASED WOLF POSITION UPDATION
for each search agent in pack p
```
update the position $\vec{X}_i^p(t)$ of the current search
agent by equation
```
          (10 to 20)
end for
 end for
%PROBABILISTIC CROSSOVER
if (rand()<Pr)
```
$\vec{X}_{child}$ =Crossover$(\vec{X}_i^k, \vec{X}_j^m)$ %SELECT ANY TWO MEMBERS FOR ANY TWO
```
PACKS k AND m
End if
%EVALUATION AND BEST WOLF SELECTION
```
if (f($\vec{X}^{\alpha pack^k}$)>f($\vec{X}_{child}$))

$\vec{X}^{\alpha pack^k} = \vec{X}_{child}$
```
end if
```
if (f($\vec{X}^{\alpha pack^m}$)>f($\vec{X}_{child}$))

$\vec{X}^{\alpha pack^m} = \vec{X}_{child}$
```
end if
%ARCHIVAL OF THE BEST WOLF
```

continued on following page

Algorithim 2. Continued

$$\left[\textit{fbest}, \textit{bestpack} \right] = \min \left\{ f\left(\vec{X}^{\alpha pack^k} \right), \left(\vec{X}^{\alpha pack^2} \right), ..., \left(\vec{X}^{\alpha pack^P} \right) \right\}$$

$$\vec{X}_{best} = \vec{X}^{\alpha pack^{bestpack}}$$

```
update a,A and C
for p=1 to P do
        for i= 1 to  M_p  do
                calculate the fitness of all search agents
        end for
        update  X^αpack^p ,  X^βpack^p ,  X^δpack^p
end for
t=t+1
end while     %END OF WHILE LOOP
return  X_best
```

algorithm is compared with other state of the problems to establish the effectiveness of the proposed algorithm.

EXPERIMENTAL SERIES 1: 23 CLASSICALBENCHMARKPROBLEMS

The description of the classical benchmark algorithms are given in Table 3, Table 4 and Table 5. We run the MMGWO algorithm 30 times for each function. The statistical results, e.g. average value (Ave) and standard deviation (STD) of 30 runs each having 500 iterations for MMGWO algorithm and other methods are shown in Table 7. The best results are marked bold in the Table 7. According to the result shown in Table 7, MMGWO gives very promising results. The algorithm performs very well for F01, F02, F03, F04, F09, F10, F11, F16, F17, F18 and F20. The results of the unimodal functions F01-F07 show the superior performance with MMGWO. Out of these seven functions we get promising results in four cases. The convergence graphs in Figure7 and barchart in Figure 4 also reflect the enhancement of the exploitation trends with unimodal function F01-F07. It provides good results compared to other basic algorithm such as Grey Wolf Optimization (GWO), Cuckoo Search (CS) (Basu & Chowdhury,2013), Particle Swarm Optimization (PSO) (Kennedy & Eberhart,1995), Firefly algorithm (FA) (Yang, 2010), Gravitational Search Algorithm (GSA), Bat algorithm (BA), Differential Evolution (DE) and GWO with Levy flight (LGWO).

Figure 3. Conceptual model of proposed algorithm

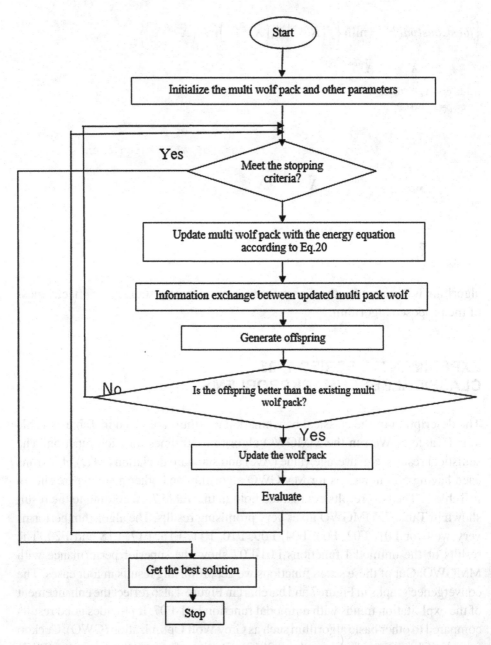

Multimodal test beds (F08-F23) are appropriate to examine the exploration tendency. According to the result shown in Table 7 and also the bar charts given in Figure 5 and Figure 6, it is observed that MMGWO provides very competitive results for F08-F23. It is noticed that the MMGWO gives superior results for functions F09, F10, F11, F16, F17, F18 and F20. Out of these sixteen multimodal functions our proposed algorithm gives good results in seven cases. Thus, it can be concluded, that as a whole the proposed algorithm provides the best result in 11 functions out of 23 functions which is shown in Table 7. The last row of Table 7 shows the rank of each algorithm with its number of the best fitness function values that solved them within the first bracket. It is evident from the table that the proposed algorithm performs the best and holds rank 1 among all other algorithms in the table. These results signify that the MMGWO algorithm has merit in terms of exploration as well as exploitation. The convergence graphs for different functions are shown in Figure 7, Figure 8 and Figure 9 and they also support the above analysis.

EXPERIMENTAL SERIES 2: CEC2014 BENCHMARKS

Here, a set of 30 functions is used to evaluate the performance of the proposed algorithm (MMGWO) with other equivalent algorithms. The dimension of the problems is set to 50. Table 8 shows the comparison between the proposed algorithm and the other seven algorithms (Slap Swarm Algorithm(SSA), Ant Lion Optimization (ALO), GWO, Sine Cosine Algorithm (SCA), Whale Optimization Algorithm(WOA), DE and CS also, Figure 15 and Figure 16 show the convergence curves for the algorithms. The results in Table 8 indicate that the proposed MMGWO gives better results than the other algorithms. The best results are marked bold in the Table 8. According to the result shown in Table 8, MMGWO gives very promising results. The algorithm outperforms for f1, f3, f4, f5, f6, f9, f10, f11, f12, f13, f14, f15, f16, f17, f19, f20, f21, f22, f24, f25, f26 and f27. The results of the unimodal functions f1-f3 show the superior performance with MMGWO. Out of these three functions we get promising results in two cases. The convergence graphs in Figure 15 (convergence graph of f1) also reflect the enhancement of the exploitation trends with unimodal function f1-f3. It provides good results compared to another basic algorithm such as SSA, ALO, GWO, SCA, WOA, DE, and CS.

Simple multimodal test beds (f4-f16) are appropriate to examine the exploration tendency. According to the result shown in Table 8 and also the bar charts given in Figure 10, Figure 11 and Figure 12 it is observed that MMGWO provides very for f4-f16. It is noticed that the MMGWO gives superior results for functions f4, f5, f6, f9, f10, f11, f12, f13, f14, f15 and f16. Out of these thirteen multimodal functions our proposed algorithm gives good results in eleven cases.

The results of the hybrid functions f17-f22 show the outperformance with MMGWO. Out of these 6 functions MMGWO gives best results for functions f17,

Table 2. Comparison of MMGWO with respect to Order of complexity

Algorithm	Initialization	Evaluate	Update	Overall
MMGWO	O(ND*iter_max)	O(cost*N*iter_max)	O(ND*iter_max)	O(FE*D+FE*cost)
GWO	O(ND*iter_max)	O(cost*N*iter_max)	O(ND*iter_max)	O(FE*D+FE*cost)
PSO	O(ND*iter_max)	O(cost*N*iter_max)	O(ND*iter_max)	O(FE*D+FE*cost)
DE	O(ND*iter_max)	O(cost*N*iter_max)	O(ND*iter_max)	O(FE*D+FE*cost)
ABC	O(ND*iter_max)	O(cost*N*iter_max)	O(ND*iter_max)	O(FE*D+FE*cost)

D=dimension of the problem, N=population size, N*iter_max=FE, cost=cost function, T=number of generations

Table 3. Unimodal benchmark functions

Function	Dim	Range	f_{min}
$F01(x) = \sum_{i=1}^{n} x_i^2$	30	[-100,100]	0
$F02(x) = \sum_{i=1}^{n} \lvert x_i \rvert + \prod_{i=1}^{n} \lvert x_i \rvert$	30	[-10,10]	0
$F03(x) = \sum_{i=1}^{n} \left(\sum_{j-1}^{i} x_j \right)^2$	30	[-100,100]	0
$F04(x) = \max_i \left\{ \lvert x_i \rvert, 1 \leq i \leq n \right\}$	30	[-100,100]	0
$F05 = \sum_{i=1}^{n-1} \left[100 \left(x_{i+1} - x_i^2 \right)^2 + \left(x_i - 1 \right)^2 \right]$	30	[-30,30]	0
$F06(x) = \sum_{i=1}^{n} \left(\lfloor x_i + 0.5 \rfloor \right)^2$	30	[-100,100]	0
$F07(x) = \sum_{i=1}^{n} i x_i^4 + random(0,1)$	30	[-1.28,1.28]	0

Table 4. Multimodal benchmark functions

Function	Dim	Range	f_{min}
$F08(x) = \sum_{i=1}^{n} -x_i \sin(\sqrt{|x_i|})$	30	[-500,500]	-418.9829 E5
$F09(x) = \sum_{i=1}^{n} \left[x_i^2 - 10\cos(2\pi x_i + 10) \right]$	30	[-5.12,5.12]	0
$F10(x)=$ $-20\exp(-0.2\sqrt{\frac{1}{n}\sum_{i=1}^{n} x_i^2} \exp(\frac{1}{n}\sum_{i=1}^{n}\cos\left(2\pi x_i\right) + 20 + e$	30	[-32,32]	0
$F11(x) = \frac{1}{400}\sum_{i=1}^{n} x_i^2 - \prod_{i=1}^{n}\cos(\frac{x}{\sqrt{i}})+1$	30	[-600,600]	0
$F12(x)=$ $\frac{\Pi}{n}\left\{ 10\sin\left(\pi\pi y_1\right) + \sum_{i=1}^{n-1}\left(y_i -1\right)^2 [1 + 10\sin^2(\pi y_{i+1})] + \left(y_n -1\right)^2 \right\}$ $+\sum_{i=1}^{nu} u(x_i,10,100,4)$ $y_i = 1 + \frac{x+1}{4}$ $u\left(x_i,a,k,m\right) = \begin{cases} k\left(x_i - a\right)^m & x_i > a \\ 0 & -a < x_i < a \\ k\left(-x_i - a\right)^m & x_i < -a \end{cases}$	30	[-50,50]	0
$F13(x)=0.1($ $\sin^2\left(3\pi x_1\right) + \sum_{i=1}^{n}[(x_i -1)^2] + (x_n -1)^2 \left[1 + \sin^2 2\left(\pi x_n\right)\right]\} + \sum_{i=1}^{n} u\left(x_i,5,100,4\right)$	30	[-50,50]	0

f19, f20, f21 and f22. The convergence graphs (Figure 16) and barcharts (Figure 12 and Figure 13) reflect the enhancement of the exploitation trends.

In case of composite functions (f23-f30) the proposed algorithm gives superior performance for functions f24, f25, f26 and f27. It is noticed that out of the eight functions, the proposed algorithm gives promising results in four cases. The convergence curve (Figure 16) and barcharts (Figure 13and Figure 14) also supports the same analysis.

Table 5. Fixed-dimension multimodal benchmark functions

Function	Dim	Range	f_{min}
$F14(x)= \left(\dfrac{1}{500} + \sum\limits_{j=1}^{25} \dfrac{1}{\sum\limits_{j=1}^{2} \left(x_i - a_{i,j} \right)^6} \right)^{-1}$	2	[-65,65]	1
$F15(x)= \sum\limits_{i=1}^{11} \left[a_i - \dfrac{x_1 \left(b_i^2 + b_i x_2 \right)}{b_i^2 + b_i x_3 + x_4} \right]^2$	4	[-5,5]	0.00030
$F16(x)= 4x_1^2 - 2.1x_1^4 + \dfrac{1}{3}x_1^6 + x_1 x_2 - 4x_2^2 + 4x_2^4$	2	[-5,5]	-1.0316
$F17(x)=$ $\left(x_2 - \dfrac{5.1}{4\pi^2}x_1^2 + \dfrac{5}{\pi}x_1 - 6 \right)^2 + 10\left(1 - \dfrac{1}{8\pi} \right)\cos x_1 + 10$	2	[-5,5]	0.398
$F18(x)=$ $\left[1 + \left(x_1 + x_2 + 1 \right)^2 \left(19 - 14x_1 + 3x_1^2 - 14x_2 + 6x_1 x_2 + 3x_2^2 \right) \right]$ $\left[30 + \left(2x_1 - 3x_2 \right)^2 E\left(18 - 32x_1 + 12x_1^2 + 48x_2 - 36x_1 x_2 + 27x_2^2 \right) \right]$	2	[-2,2]	3
$F19(x)= -\sum\limits_{i=1}^{4} c_i \exp\left(-\sum\limits_{j=1}^{3} a_{i,j} \left(x_j - p_{i,j} \right)^2 \right)$	3	[1,3]	-3.86
$F20(x)= -\sum\limits_{i=1}^{4} c_i \exp\left(-\sum\limits_{j=1}^{6} a_{i,j} \left(x_j - p_{i,j} \right)^2 \right)$	6	[0,1]	-3.32
$F21(x)= -\sum\limits_{i=1}^{5} \left[\left(X - a_i \right)\left(X - a_i \right)^T + c_i \right]^{-1}$	4	[0,10]	-10.1532
$F22(x)= -\sum\limits_{i=1}^{7} \left[\left(X - a_i \right)\left(X - a_i \right)^T + c_i \right]^{-1}$	4	[0,10]	-10.4028
$F23(x)= -\sum\limits_{i=1}^{10} \left[\left(X - a_i \right)\left(X - a_i \right)^T + c_i \right]^{-1}$	4	[0,10]	-10.5363

Table 6. The definition of CEC2014 Benchmark

	No.	Functions	$F_i^* = F_i(x^*)$
Unimodal Functions	1	Rotated High conditioned Elliptic Function	100
	2	Rotated Bent Ciger function	200
	3	Rotated Discus function	300
Simple Multimodal Functions	4	Shifted and Rotated Rosenbrock's Function	400
	5	Shifted and Rotated Ackley's Function	500
	6	Shifted and Rotated Weierstrass Function	600
	7	Shifted and Rotated Griewank's Function	700
	8	Shifted Rastrigin's Function	800
	9	Shifted and Rotated Rastrigin's Function	900
	10	Shifted Schwefel's Function	1000
	11	Shifted and Rotated Schwefel's Function	1100
	12	Shifted and Rotated Katsuura Function	1200
	13	Shifted and Rotated HappyCat Function	1300
	14	Shifted and Rotated HGBat Function	1400
	15	Shifted and Rotated Expanded Griewank's plus Rosenbrock's Function	1500
	16	Shifted and Rotated Expanded Scaffer's F_6 Function	1600
Hybrid Function 1	17	Hybrid Function 1 (*N=3*)	1700
	18	Hybrid Function 2 (*N=3*)	1800
	19	Hybrid Function 3 (*N=4*)	1900
	20	Hybrid Function 4 (*N=4*)	2000
	21	Hybrid Function 5 (*N=5*)	2100
	22	Hybrid Function 6 (*N=5*)	2200
Composition Functions	23	Composition Functions 1 (*N=5*)	2300
	24	Composition Functions 2 (*N=3*)	2400
	25	Composition Functions 3 (*N=3*)	2500
	26	Composition Functions 4 (*N=5*)	2600
	27	Composition Functions 5 (*N=5*)	2700
	28	Composition Functions 6 (*N=5*)	2800
	29	Composition Functions 7 (*N=3*)	2900
	30	Composition Functions 8 (*N=3*)	3000

Table 7. The results of average fitness values and STD values for30-dimensional classical benchmark problems [10] for 500 iterations

		GWO	CS	PSO	FA	GSA	BA	DE	LGWO	MMGWO
F01	Ave	6.32E-28	5.78E-03	0.000128	0.040411	2.12E-16	0.767411	7.17E-12	3.17E-30	**0.00E+00**
	STD	5.18E-05	2.41E-03	0.000289	0.018201	5.21E-17	0.68817	2.10E-14	4.07E-20	**0.00E+00**
F02	Ave	8.38E-17	2.08E-01	0.048122	0.061744	0.051774	0.32711	1.10E-07	5.39E-19	**0.00E+00**
	STD	0.034082	3.17E-02	0.041385	0.017173	0.231239	2.160233	3.35E-08	0.010729	**0.00E+00**
F03	Ave	4.09E-06	2.63E-01	64.17332	0.050016	455.1004	0.129385	8.23E-07	8.12E-08	**0.00E+00**
	STD	21.70012	2.97E-02	18.7181	0.01125	124.7488	0.688058	2.31E-07	2.053381	**0.00E+00**
F04	Ave	8.19E-07	1.43E-05	1.38855	0.213711	8.021015	0.193882	1.18E-04	1.17E-08	**0.00E+00**
	STD	1.741844	4.83E-06	0.325442	0.028923	1.49311	0.641056	1.12E-05	1.316448	**0.00E+00**
F05	Ave	24.74168	0.008121	85.03224	3.005223	58.3892	0.32171	**0**	8.350714	28.095
	STD	51.82238	0.054326	50.40314	1.563822	56.00025	0.299057	0.00E+00	5.336001	**4.41E-01**
F06	Ave	0.798003	6.17E-04	0.000425	0.04523	**8.31E-14**	0.712085	2.88E-03	2.69E-04	0.031703
	STD	0.001289	2.80E-05	1.27E-05	0.029208	7.70E-15	0.892204	1.45E-05	0.000023	**8.38E-02**
F07	Ave	0.003781	0.028551	0.102744	0.008561	0.088145	0.153102	2.41E-02	**3.02E-03**	0.101013
	STD	0.214005	0.001277	0.04235	0.004512	0.044009	0.10873	1.12E-03	**0.001102**	5.68E-02
F08	Ave	−4021.351	−2128.913	−4258.206	−1295.180	−2688.746	−1002.711	−1538.1527	−3365.8658	-4.57E+03
	STD	315.447	0.008418	512.418	302.4846	475.2366	831.0054	582.4522	296.12698	1.12E+03
F09	Ave	3.03E-01	0.246322	34.18946	0.2833	21.56339	1.200358	12.414778	0.094579	**0**
	STD	25.26322	0.054326	8.296435	0.211472	5.1287386	0.642822	9.2482451	21.580073	**0.00E+00**
F10	Ave	1.12E-13	4.01E-10	0.258511	0.142184	0.096552	0.130025	2.85E-10	2.12E-15	**8.88E-16**
	STD	0.085041	5.21E-09	0.489455	0.042488	0.195096	0.064851	3.10E-08	0.0429752	**2.04E-31**
F11	Ave	0.005102	0.185228	0.008922	0.086059	12.33364	1.138114	8.15E-04	2.42E-05	**0**
	STD	0.006325	0.039805	0.009626	0.028002	4.801125	0.637005	1.41E-06	0.0000839	**0.00E+00**
F12	Ave	0.061741	0.012581	0.007241	0.131795	2.112974	0.409255	8.21E-03	**7.12E-04**	9.43E-03
	STD	0.006325	0.039805	0.009626	0.028002	4.801125	0.637005	**1.41E-06**	0.0000839	8.54E-03
F13	Ave	0.512798	0.485117	0.017158	0.002966	4.200091	0.362611	5.15E-06	**3.94E-07**	0.056343
	STD	0.005821	**6.85E-08**	0.018251	0.001452	2.911045	0.189552	1.52E-07	0.0002098	4.67E-02
F14	Ave	4.013365	1.423652	3.836673	3.026592	4.980711	3.500296	9.95E-01	1.149379	26.56401
	STD	4.030052	1.30E-02	3.100205	1.932225	3.955225	2.305466	**2.56E-11**	2.9072365	8.45E+01
F15	Ave	0.0005837	5.03E-04	0.000612	0.001	0.003569	7.03E-03	5.48E-05	**2.53E-06**	5.26E-04
	STD	0.0007215	1.11E-04	0.000198	4.60E-04	0.001798	2.11E-03	**1.18E-06**	3.95E-04	1.57E-04
F16	Ave	−1.03163	−1.03163	−1.03163	−1.03160	−1.03163	−1.03163	−1.03163	−1.03163	**-1.0316**
	STD	4.34E+00	1.49E-08	**6.27E-16**	1.65E-07	4.72E-16	2.452254	3.17E-11	3.20E-12	7.30E-05
F17	Ave	0.397889	0.39795	0.397901	0.397985	0.397887	0.3979	3.98E-01	0.397887	**0.3979**
	STD	1.28E-05	3.24E-06	**0**	3.56E-08	0.00E+00	1.27E-02	8.24E-07	0.00E+00	3.79E-05
F18	Ave	3.000028	3.00135	3.0001	3.012363	3	3	3.0000E+00	3	**3.00E+00**
	STD	0.012257	0.00258	7.01E-06	0.052675	4.29E-14	0.056167	**1.58E-18**	2.40E-02	2.15E-03
F19	Ave	−3.86263	−3.86288	−3.86280	−3.86137	−3.86250	−3.86280	−3.86280	**−3.86288**	-0.3005
	STD	2.122054	1.85E-05	1.89E-12	3.81E-03	3.15E-15	1.104492	1.51E-20	1.47E-05	**0.00E+00**

continued on following page

Table 7. Continued

		GWO	CS	PSO	FA	GSA	BA	DE	LGWO	MMGWO
F20	Ave	−3.28635	−3.32185	−3.26632	−3.28415	−3.31862	−3.322084	−3.2176	−3.32101	**-3.17559**
	STD	0.268112	7.21E-03	0.060413	0.070285	**4.15E-05**	0.058045	0.63565	0.038626	0.176
F21	Ave	−10.1510	−9.72828	−7.63838	−6.92669	−5.92533	−9.203955	−10.1527	**−10.15304**	-7.73
	STD	9.120014	0.288102	3.672215	3.257075	3.730022	3.0581205	**6.53E-05**	5.14E+00	2.86
F22	Ave	−10.255746	−9.87298	−7.36982	−10.4008	−9.630195	−9.23499	−10.4028	**−10.4028**	7.4093
	STD	8.695223	**0.320344**	3.255826	1.198224	2.8100546	3.225782	2.625526	2.200478	1.9936
F23	Ave	−10.53435	−9.78223	−9.968005	−10.2182	−10.95211	−10.98552	−10.5364	**−10.53640**	-8.72
	STD	8.4924069	0.500213	1.98051	1.240585	**1.41E-14**	10.29354	1.055221	2.0122878	3.40092
Rank		4(0)	4(0)	4(0)	4(0)	3(1)	4(0)	3(1)	2(8)	**1(11)**

Thus, it can be concluded, that as a whole the proposed algorithm provides the best result in 22 functions out of 30 functions which is shown in Table 8. The last row of Table 8 shows the rank of each algorithm with its number of the best fitness function values that solved them within the first bracket. It is evident from the table that the proposed algorithm performs the best and holds rank 1 among all other algorithms in the table. These results signify that the MMGWO algorithm has merit in terms of exploration as well as exploitation. The convergence graphs for different functions are shown in Figure 15 and Figure 16 and they also support the above analysis.

EXPERIMENTAL SERIES 3: MMGWO FOR REAL TIME ENGINEERING PROBLEMS

Pressure Vessel Design Problem

In this section a constrained engineering design problem entitled as pressure vessel design is analyzed with the proposed MMGWO algorithm. Here the main aim of this problem is to optimize the total cost of material and welding of a cylindrical vessel. There are four variables to be optimized in this problem and they are-thickness of the shell (T_s); thickness of the head (T_h); inner radius(R) and length of the cylindrical portion without considering head (L). This engineering problem is subjected to four constraints. Here we considered $\vec{x} = [\, x_1 x_2 x_3 x_4 \,] = [\text{T}_s \text{T}_h \text{RL}]$,

$$\text{Minimize } f\left(\vec{x}\right) = 0.6224\, x_1 x_3 x_4 + 1.7781\, x_2 x_3^2 + 3.1661\, x_1^2 x_4 + 19.84\, x_1^2 x_3$$

$$\text{subject to } g_1\left(\vec{x}\right) = -x_1 + 0.0193 x_3 \leq 0,$$

Figure 4. Barchart for the comparison of different algorithms for unimodal functions (F01-F07)

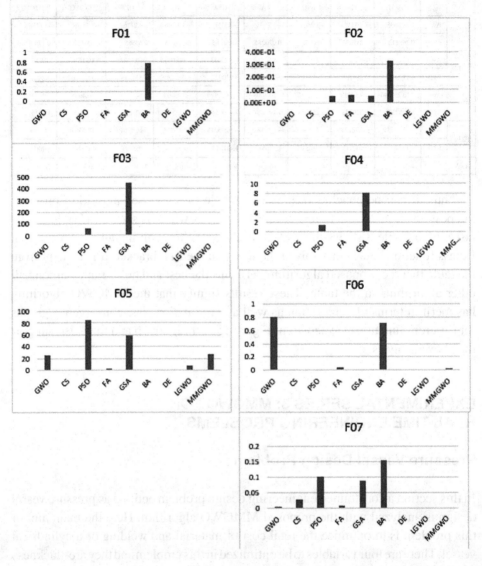

$$g_2(\vec{x}) = -x_3 + 0.00954x_3 \leq 0,$$

$$g_3(\vec{x}) = -\Pi x_3^2 - \frac{4^3}{3^3} + 1296000 \leq 0,$$

$$g_4(\vec{x}) = x_4 - 240 \leq 0$$

Variable range:

$$0 \leq x_1 \leq 99$$

Figure 5. Barchart for the comparison of different algorithms for multimodal functions (F07-F15)

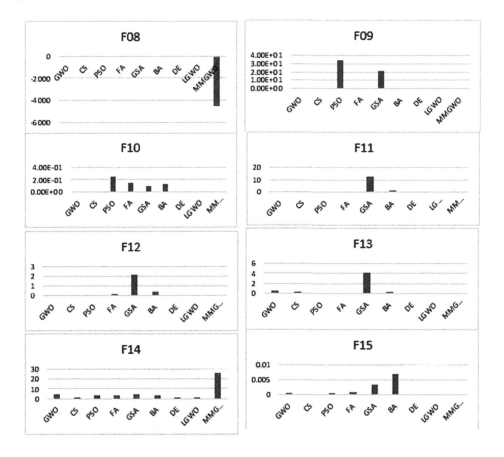

Figure 6. Barchart for the comparison of different algorithms for multimodal functions (F16-F23)

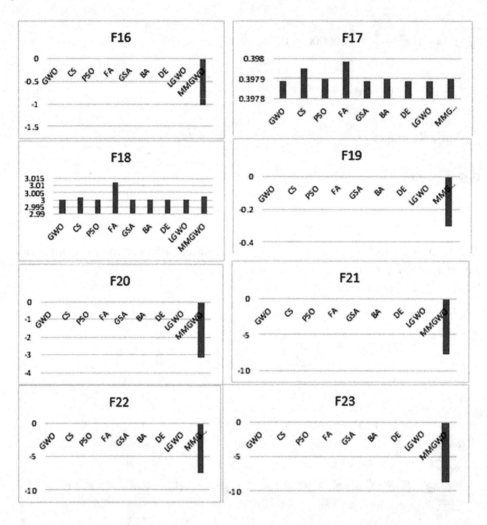

$$0 \leq x_2 \leq 99$$

$$10 \leq x_3 \leq 200$$

$$10 \leq x_{34} \leq 200$$

From the above table, we conclude that the proposed algorithm gives better performance than conventional GWO, PSO, GA and ES algorithm.

Figure 7. Convergence curve for different unimodal functions (F01-F04)

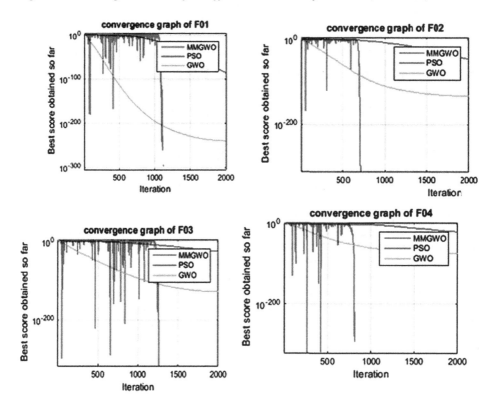

Welded Beam Design

The objective of this problem is to minimize the fabrication cost of a welded beam as shown in Figure.18. The constraints are: shear stress (τ), bending stress in the beam (θ), buckling load on the bar (P_c), end deflection of the beam (δ), side constraints. Here we consider

$$\vec{x} = \left[x_1 x_2 x_3 x_4 \right] = \left[h l t b \right],$$

Minimize $f(\vec{x}) = 1.10471 x_1^2 + 0.04811 x_3 x_4 \left(14.0 + x_2 \right).$

Subject to

$$g_1(\vec{x}) = \tau(\vec{x}) - \tau_{\max} \leq 0$$

Figure 8. Convergence curve for different multimodal functions (F09, F10, F11, F17, F18)

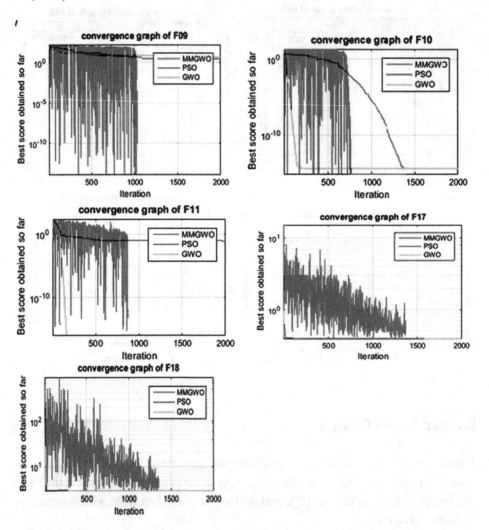

$$g_2(\overrightarrow{x}) = \sigma(\overrightarrow{x}) - \sigma_{max} \leq 0$$

$$g_3(\overrightarrow{x}) = \delta(\overrightarrow{x}) - \delta_{max} \leq 0$$

$$g_4(\overrightarrow{x}) = x_1 - x_4 \leq 0$$

$$g_5(\overrightarrow{x}) = P - P_c(\overrightarrow{x}) \leq 0$$

$$g_6(\overrightarrow{x}) = 0.125 - x_1 \leq 0$$

$$g_7(\overrightarrow{x}) = 1.10471x_1 + 0.04811x_3x_4(14.0 + x_2) - 5.0 \leq 0$$

Variable range

$$0.1 \leq x_1 \leq 2,$$

$$0.1 \leq x_2 \leq 10,$$

$$0.1 \leq x_3 \leq 10,$$

$$0.1 \leq x_4 \leq 2.$$

Where

Figure 9. Convergence curve for multimodal function (F20)

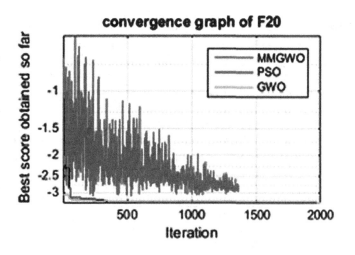

Table 8. The results of average fitness values and STD values for CEC2014 benchmark functions for 5000000 Functions Evaluation (FE)

		SSA	ALO	GWO	SCA	WOA	DE	CS	MMGWO
f1	Ave	4.69E+07	2.84E+07	1.42E+08	1.18E+09	3.10E+08	1.85E+10	6.02E+08	**2.07E+07**
	STD	1.78E+07	1.00E+07	6.31E+07	3.04E+09	9.03E+07	2.99E+09	1.01E+08	7.78E+06
f2	Ave	8.69E+03	2.26E+04	1.14E+10	7.62E+10	1.51E+10	3.51E+08	2.28E+08	8.83E+07
	STD	1.01E+04	1.16E+04	5.10E+09	9.81E+09	3.11E+09	3.64E+08	1.18E+08	4.40E+07
f3	Ave	1.29E+05	1.72E+05	1.14E+10	1.38E+05	1.47E+05	8.46E+05	7.11E+05	**4.93E+04**
	STD	2.72E+04	4.42E+04	2.03E+04	1.85E+04	2.44E+04	1.71E+05	1.30E+05	1.21E+04
f4	Ave	5.92E+02	6.59E+02	1.73E+03	1.38E+04	2.37E+03	6.87E+02	6.54E+02	**5.32E+02**
	STD	7.63E+01	8.11E+01	8.37E+02	3.05E+03	6.58E+02	2.96E+02	8.21E+01	3.39E+01
f5	Ave	5.20E+02	5.20E+02	5.21E+02	5.21E+02	5.21E+02	5.21E+02	5.21E+02	**5.20E+02**
	STD	9.75E-02	1.01E-01	4.01E-02	3.25E-02	7.69E-02	4.02E-02	3.60E-02	1.58E-01
f6	Ave	6.49E+02	6.53E+02	6.34E+02	6.71E+02	6.70E+02	6.69E+02	6.54E+02	**6.30E+02**
	STD	5.35E+00	4.02E+00	3.70E+00	2.69E+00	3.17E+00	2.08E+00	2.06E+00	3.92E+00
f7	Ave	7.00E+02	7.00E+02	8.18E+02	1.40E+03	8.18E+02	7.00E+02	7.03E+02	7.02E+02
	STD	1.78E-02	1.01E-01	4.79E+01	8.60E+01	3.71E+01	9.22E-02	1.03E+00	4.93E-01
f8	Ave	1.10E+03	1.08E+03	1.03E+03	1.38E+03	1.22E+03	**1.01E+03**	1.15E+03	1.05E+03
	STD	5.32E+01	4.72E+01	3.81E+01	3.39E+01	4.31E+01	1.19E+01	4.70E+01	2.28E+01
f9	Ave	1.22E+03	1.22E+03	1.14E+03	1.51E+03	1.48E+03	1.31E+03	1.42E+03	**1.14E+03**
	STD	5.32E+01	4.72E+01	3.81E+01	3.39E+01	4.31E+01	1.19E+01	4.70E+01	2.50E+01
f10	Ave	7.87E+03	8.00E+03	7.20E+03	1.43E+04	1.11E+04	9.65E+03	1.29E+04	**6.64E+03**
	STD	8.40E+02	1.07E+03	1.04E+03	5.73E+02	1.21E+03	3.82E+02	5.10E+02	3.87E+02
f11	Ave	8.07E+03	8.75E+03	9.04E+03	1.53E+04	1.28E+04	1.66E+04	1.55E+04	**6.59E+03**
	STD	8.94E+02	1.07E+03	2.83E+03	4.43E+02	1.07E+03	4.10E+02	4.37E+02	6.56E+02
f12	Ave	**1.20E+03**	**1.20E+03**	**1.20E+03**	**1.20E+03**	**1.20E+03**	**1.20E+03**	**1.20E+03**	**1.20E+03**
	STD	3.90E-01	4.27E-01	1.58E+00	3.90E-01	5.46E-01	3.23E-01	3.15E-01	2.76E-01
f13	Ave	**1.30E+03**	**1.30E+03**	**1.30E+03**	**1.30E+03**	**1.30E+03**	**1.30E+03**	**1.30E+03**	**1.30E+03**
	STD	1.18E-01	1.04E-01	8.87E-01	3.40E-01	7.77E-01	7.37E-02	8.36E-02	1.03E-01
f14	Ave	**1.40E+03**	**1.40E+03**	1.42E+03	1.57E+03	1.42E+03	**1.40E+03**	**1.40E+03**	**1.40E+03**
	STD	2.31E-01	3.90E-02	1.35E+01	1.84E+01	6.84E+00	1.05E-01	1.21E-01	2.94E-01
f15	Ave	**1.54E+03**	1.55E+03	5.46E+03	4.77E+05	3.37E+04	1.60E+03	1.76E+03	**1.51E+03**
	STD	1.11E+01	1.42E+01	3.64E+03	1.83E+05	2.84E+04	5.17E+01	2.71E+02	3.04E+00
f16	Ave	**1.62E+03**	**1.62E+03**	**1.62E+03**	**1.62E+03**	**1.62E+03**	**1.62E+03**	**1.62E+03**	**1.62E+03**
	STD	7.32E-01	6.09E-01	9.59E-01	2.90E-01	5.40E-01	1.08E-01	1.48E-01	8.06E-01
f17	Ave	3.65E+06	4.23E+06	9.49E+06	9.48E+07	1.65E+08	9.48E+08	5.11E+07	**1.54E+06**
	STD	2.12E+06	2.48E+06	6.95E+06	3.55E+07	9.69E+07	2.63E+08	1.47E+07	9.76E+05

continued on following page

Table 8. Continued

		SSA	ALO	GWO	SCA	WOA	DE	CS	MMGWO
f18	Ave	1.70E+10	**3.94E+03**	1.25E+08	2.56E+09	5.73E+07	5.24E+07	8.39E+05	7.63E+04
	STD	1.57E+03	1.13E+03	1.77E+08	6.75E+08	8.91E+07	5.48E+07	6.99E+05	1.40E+05
f19	Ave	1.96E+03	1.97E+03	2.01E+03	2.29E+03	2.15E+03	1.95E+03	1.94E+03	**1.93E+03**
	STD	1.67E+01	2.87E+01	4.18E+01	5.54E+01	7.51E+01	3.07E+01	3.51E+00	5.09E+00
f20	Ave	6.07E+04	1.18E+05	3.99E+04	7.84E+04	6.47E+05	1.39E+06	2.29E+05	**3.10E+04**
	STD	2.25E+04	4.10E+04	1.17E+04	3.14E+04	6.56E+05	3.96E+05	9.33E+04	7.46E+03
f21	Ave	2.41E+06	2.72E+06	4.83E+06	1.07E+08	2.11E+07	3.97E+08	1.05E+07	**4.83E+05**
	STD	1.34E+06	1.52E+06	3.81E+06	6.68E+06	1.17E+07	1.51E+08	3.69E+06	2.96E+05
f22	Ave	3.51E+03	3.84E+03	3.19E+03	5.08E+03	4.84E+03	4.31E+03	4.16E+03	**2.95E+03**
	STD	2.60E+02	3.80E+02	2.84E+02	3.36E+02	7.00E+02	1.89E+02	1.54E+02	2.89E+02
f23	Ave	2.69E+03	2.68E+03	2.77E+03	3.19E+03	2.83E+03	2.64E+03	2.64E+03	2.67E+03
	STD	1.29E+01	1.03E+01	5.02E+01	1.11E+02	1.45E+02	7.00E-01	1.67E+00	5.96E+00
f24	Ave	2.70E+03	2.74E+03	2.60E+03	2.75E+03	2.60E+03	2.67E+03	2.69E+03	**2.60E+03**
	STD	1.20E+01	1.96E+01	1.94E-02	4.15E+01	2.99E+00	1.13E+00	5.58E+00	4.66E-02
f25	Ave	2.74E+03	2.76E+03	2.72E+03	2.79E+03	2.70E+03	4.36E+03	2.76E+03	**2.70E+03**
	STD	7.92E+00	1.13E+01	1.69E+01	2.54E+01	1.81E+01	2.84E+02	1.20E+01	0.00E+00
f26	Ave	2.71E+03	2.74E+03	2.80E+03	2.71E+03	2.71E+03	**2.70E+03**	**2.70E+03**	**2.70E+03**
	STD	5.88E+01	5.17E+01	4.52E+01	7.47E-01	3.00E+01	9.20E-02	8.69E-02	8.99E-02
f27	Ave	4.29E+03	4.40E+03	3.93E+03	4.95E+03	4.91E+03	4.78E+03	4.41E+03	**3.49E+03**
	STD	1.54E+02	1.40E+02	1.22E+02	7.72E+01	1.42E+02	4.40E+01	2.81E+01	7.06E+01
f28	Ave	5.91E+03	8.49E+03	5.98E+03	1.04E+04	9.44E+03	2.30E+04	3.75E+03	4.21E+03
	STD	8.17E+02	9.92E+02	7.92E+02	9.74E+02	1.89E+03	3.69E+03	1.23E+02	1.13E+02
f29	Ave	5.39E+07	4.89E+07	1.47E+07	3.19E+08	8.30E+07	3.54E+03	3.21E+03	1.60E+04
	STD	5.27E+07	2.67E+08	1.37E+07	4.98E+07	5.56E+07	1.31E+02	2.57E+01	6.56E+03
f30	Ave	8.07E+04	1.01E+05	3.38E+05	4.46E+06	1.13E+06	8.42E+03	5.53E+03	1.23E+05
	STD	2.87E+04	3.52E+04	1.87E+05	2.01E+06	1.06E+06	8.99E+02	2.55E+02	5.50E+04
Rank		4(4)	3(5)	5(3)	5(3)	5(3)	2(6)	3(5)	**1(22)**

Figure 10. Barchart of the CEC2014 functions (f1-f6) for the comparison of different algorithm

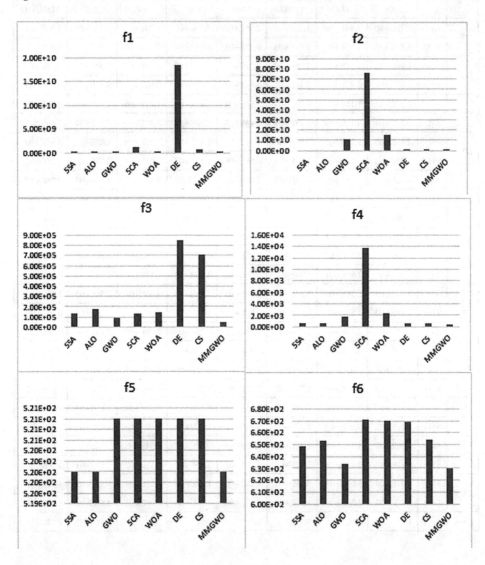

$$\tau\left(\vec{x}\right) = \sqrt{\left(\tau'\right)^2 + 2\tau'\tau''\frac{x_2}{2R} + \left(\tau''\right)^2}$$

$$\tau' = \frac{P}{\sqrt{2}x_1x_2}, \tau'' = \frac{MR}{J}, M = P\left(L + \frac{x_2}{2}\right),$$

Figure 11. Barchart of the CEC2014 functions (f7-f12) for the comparison of different algorithms

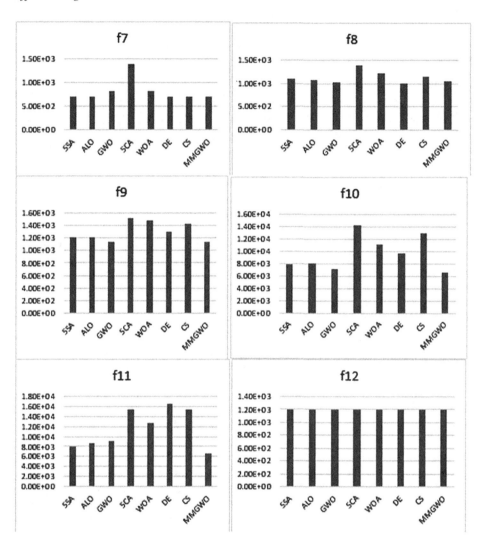

$$R = \sqrt{\frac{x_2^2}{4} + \left(\frac{x_1 + x_3}{2}\right)^2} \, ,$$

$$J = 2\left\{\sqrt{2}x_1 x_2 \left[\frac{x_2^2}{2} + \left(\frac{x_1 + x_3}{2}\right)^2\right]\right\}$$

Figure 12. Barchart of the CEC2014 functions (f13-f18) for the comparison of different algorithms

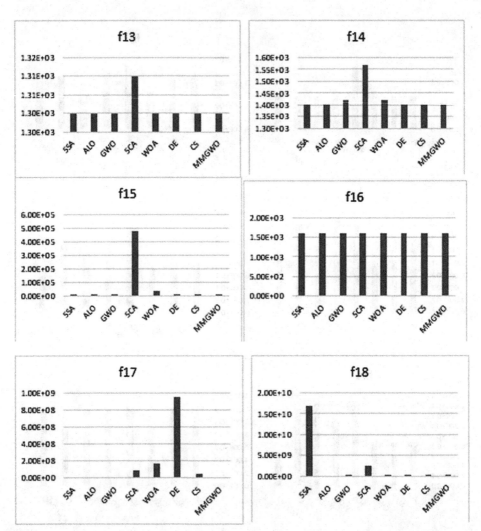

$$\sigma(\vec{x}) = \frac{6PL}{x_4 x_3^2} \ , \ \delta(\vec{x}) = \frac{6PL^3}{Ex_3^2 x_4}$$

$$P_c(\vec{x}) = \frac{4.013E\sqrt{\dfrac{x_3^2 x_4^6}{36}}}{L^2} \left(1 - \frac{x_3}{2L}\sqrt{\frac{E}{4G}}\right),$$

Figure 13. Barchart of the CEC2014 functions (f19-f24) for the comparison of different algorithms

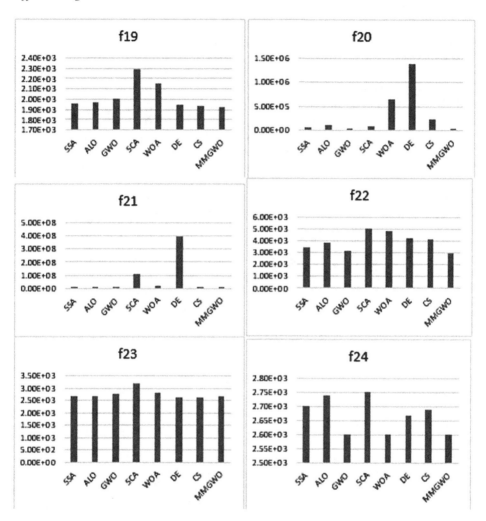

$$P = 6000lb, L = 14in., \delta_{max} = 0.25in., E = 3 \times 10^6 \text{psi},$$

$$G = 12 \times 10^6 \text{ psi}, \tau_{max} = 13600 \text{ psi}, \sigma_{max} = 30000 \text{ psi}$$

From Table 10, we conclude that the proposed algorithm gives better performance than conventional GWO, GSA, GA (Coello) and GA (Deb) algorithm.

Figure 14. Barchart of the CEC2014 functions (f25-f30) for the comparison of different algorithms

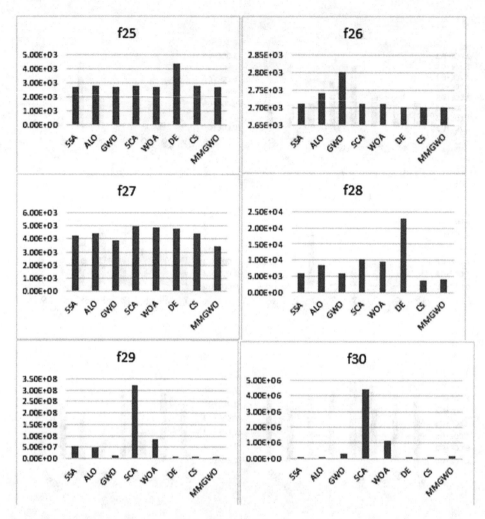

FUTURE WORK AND CONCLUSION

In this chapter, a new concept of the modified multi grey wolf pack structure is introduced for global optimization. This can act as an effective support for decision making in any disease diagnosis system and very much applicable for rural healthcare. The introduction of multi wolf pack structure, dividing the entire population into multiple packs, modification of learning technique, information sharing between the multi wolf pack populations help to manage the exploration and exploitation in a balanced way. The proposed work is evaluated for the global optimization. The

Figure 15. Convergence graph of the CEC2014 functions (f1, f4, f5, f6)

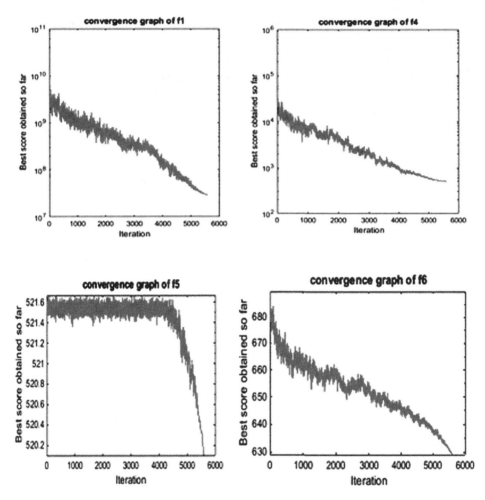

comparative studies establish the efficiency, robustness and converging property of MMGWO.

However, the limitation of the proposed algorithm is discussed in this paragraph. In the proposed modification, in natural course the number of parameters have been increased a few rather than the basic one. The simultaneous and optimal selection of the entire set of parameters becomes an involved task and we need to introduce further modification with minimal set of parameters. The algorithm has been tested here with two packs but further analysis is required and we need to test the effect of increasing the number of pack on system exploration and exploitation and system complexity. Also the conflict and mutual information sharing among the packs are required to be analyzed at greater extent.

Figure 16. Convergence graph of the CEC2014 functions (f9, f17, f19 and f27)

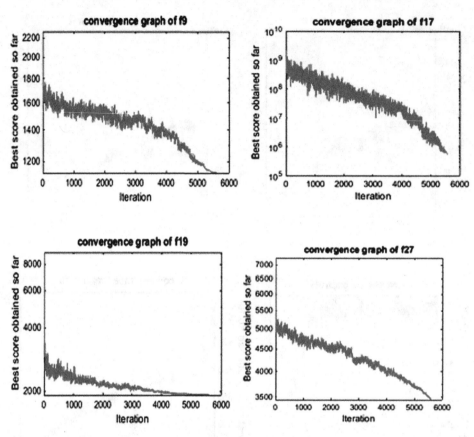

Figure 17. Pressure vessel schematic diagram

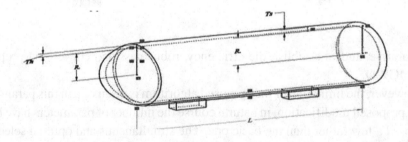

In near future, we plan to apply the proposed technique for the development of vital sign-based disease diagnosis system. The basic vital signs like heart rate, respiration rate, pulse rate, lung water content etc may be used as an input for the diagnosis system. The corresponding signals against each vital sign will provide

Table 9. The results for pressure vessel design problem

Algorithm	T_s	T_h	R	L	Optimum Cost
MMGWO	0.7887	0.3911	40.8554	192.6795	**5908.5**
GWO	0.8125	0.4345	42.0891	176.7587	6051.564
PSO	0.8125	0.4375	42.0912	176.7465	6061.078
GA	0.8125	0.4345	40.3239	200	6288.745
ES	0.8125	0.4375	42.0980	176.6405	6059.746

Figure 18. Structure of Welded beam design schematic diagram

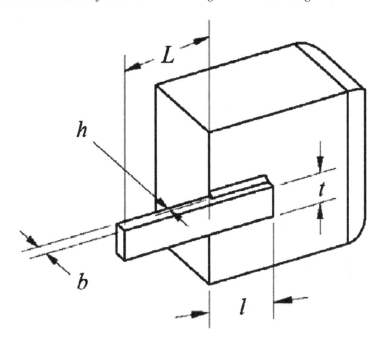

Table 10. The results for welded beam design problem

Algorithms	h	l	t	b	Optimum Cost
MMGWO	0.20609	3.4748	9.0271	0.20649	**1.7365**
GWO	0.19983	3.8064	8.7891	0.21789	1.8085
GSA	0.182129	3.856979	10.00000	0.202376	1.879952
GA(Coello)	N/A	N/A	N/A	N/A	1.8245
GA(Deb)	N/A	N/A	N/A	N/A	2.3800

information rich unique feature. The extracted features are fed to the decision support system and corresponding disease related information is extracted.

REFERENCES

Agarwalla, P., & Mukhopadhyay, S. (2020). Hybrid advanced player selection strategy based population search for global optimization. *Expert Systems with Applications*, *139*, 112825. doi:10.1016/j.eswa.2019.112825

Al-Betar, M. A., Awadallah, M. A., Faris, H., Aljarah, I., & Hammouri, A. I. (2018). Natural selection methods for grey wolf optimizer. *Expert Systems with Applications*, *113*, 481–498. doi:10.1016/j.eswa.2018.07.022

Azizi, R. (2014). *Empirical study of artificial fish swarm algorithm.* arXiv preprint arXiv:1405.4138

Banati, H., & Chaudhary, R. (2017). Multi-modal bat algorithm with improved search (MMBAIS). *Journal of Computational Science*, *23*, 130–144. doi:10.1016/j.jocs.2016.12.003

Bansal, J. C., Sharma, H., Jadon, S. S., & Clerc, M. (2014). Spider monkey optimization algorithm for numerical optimization. *Memetic Computing*, *6*(1), 31-47.

Basu, M., & Chowdhury, A. (2013). Cuckoo search algorithm for economic dispatch. *Energy*, *60*, 99–108. doi:10.1016/j.energy.2013.07.011

Bonabeau, E., Marco, D. D. R. D. F., Dorigo, M., & Theraulaz, G. (1999). *Swarm intelligence: from natural to artificial systems (No. 1).* Oxford University Press.

Celik, Y., & Ulker, E. (2013). An improved marriage in honey bees optimization algorithm for single objective unconstrained optimization. *The Scientific World Journal*. PMID:23935416

Cheema, A., & Singh, M. (2019, April). An application of phonocardiography signals for psychological stress detection using non-linear entropy-based features in empirical mode decomposition domain. *Applied Soft Computing*, *77*, 24–33. doi:10.1016/j.asoc.2019.01.006

Chen, Y., Li, L., Peng, H., Xiao, J., Yang, Y., & Shi, Y. (2017). Particle swarm optimizer with two differential mutation. *Applied Soft Computing*, *61*, 314–330. doi:10.1016/j.asoc.2017.07.020

Dilmac, S., & Korurek, M. (2015). ECG heart beat classification method based on modified ABC algorithm. *Applied Soft Computing*, *36*, 641–655. doi:10.1016/j.asoc.2015.07.010

Dorigo, M., Birattari, M., & Stutzle, T. (2006). Ant colony optimization. *Comput Intell Magaz, IEEE*, *1*(4), 28–39. doi:10.1109/MCI.2006.329691

Eberhart, R., & Kennedy, J. (1995, November). Particle swarm optimization. In *Proceedings of the IEEE international conference on neural networks* (Vol. 4, pp. 1942-1948). IEEE. 10.1109/ICNN.1995.488968

Emary, E., Zawbaa, H. M., & Hassanien, A. E. (2016). Binary grey wolf optimization approaches for feature selection. *Neurocomputing*, *172*, 371–381. doi:10.1016/j.neucom.2015.06.083

Gandomi, A. H., & Yang, X.-S. (2014). Chaotic bat algorithm. *Journal of Computational Science*, *5*(2), 224–232. doi:10.1016/j.jocs.2013.10.002

Heidari, A. A., & Pahlavani, P. (2017). An efficient modified grey wolf optimizer with Lévy flight for optimization tasks. *Applied Soft Computing*, *60*, 115–134. doi:10.1016/j.asoc.2017.06.044

Jadhav, A. N., & Gomathi, N. (2018). WGC: hybridization of exponential grey wolf optimizer with whale optimization for data clustering. *Alexandria Engineering Journal, 57*(3), 1569-1584.

Jayabarathi, T., Raghunathan, T., Adarsh, B. R., & Suganthan, P. N. (2016). Economic dispatch using hybrid grey wolf optimizer. *Energy*, *111*, 630–641. doi:10.1016/j.energy.2016.05.105

Kachuee, M., Kiani, M. M., Mohammadzade, H., & Shabany, M. (2015, May). Cuff-less high-accuracy calibration-free blood pressure estimation using pulse transit time. In 2015 IEEE international symposium on circuits and systems (ISCAS) (pp. 1006-1009). IEEE. doi:10.1109/ISCAS.2015.7168806

Karaboga, D., & Basturk, B. (2007). A powerful and efficient algorithm for numerical function optimization: Artificial bee colony (ABC) algorithm. *Journal of Global Optimization*, *39*(3), 459–471. doi:10.100710898-007-9149-x

Kumar, V., & Kumar, D. (2017). An astrophysics-inspired grey wolf algorithm for numerical optimization and its application to engineering design problems. *Advances in Engineering Software*, *112*, 231–254. doi:10.1016/j.advengsoft.2017.05.008

Lahmiri, S., & Shmuel, A. (2019). Performance of machine learning methods applied to structural MRI and ADAS cognitive scores in diagnosing Alzheimer's disease. *Biomedical Signal Processing and Control*, *52*, 414–419. doi:10.1016/j.bspc.2018.08.009

Li, Q., Chen, H., Huang, H., Zhao, X., Cai, Z., Tong, C., & Tian, X. (2017). An enhanced grey wolf optimization-based feature selection wrapped kernel extreme learning machine for medical diagnosis. *Computational and Mathematical Methods in Medicine*. PMID:28246543

Liao, W. H., Kao, Y., & Li, Y. S. (2011). A sensor deployment approach using glowworm swarm optimization algorithm in wireless sensor networks. *Expert Systems with Applications*, *38*(10), 12180–12188. doi:10.1016/j.eswa.2011.03.053

Lu, C., Gao, L., Li, X., & Xiao, S. (2017). A hybrid multi-objective grey wolf optimizer for dynamic scheduling in a real-world welding industry. *Engineering Applications of Artificial Intelligence*, *57*, 61–79. doi:10.1016/j.engappai.2016.10.013

Lu, C., Gao, L., & Yi, J. (2018). Grey wolf optimizer with cellular topological structure. *Expert Systems with Applications*, *107*, 89–114. doi:10.1016/j.eswa.2018.04.012

Lynn, N., & Suganthan, P. N. (2017). Ensemble particle swarm optimizer. *Applied Soft Computing*, *55*, 533–548. doi:10.1016/j.asoc.2017.02.007

Mech, L. D. (1999). Alpha status, dominance, and division of labor in wolf packs. *Canadian Journal of Zoology*, *77*(8), 1196–1203. doi:10.1139/z99-099

Medjahed, S. A., Saadi, T. A., Benyettou, A., & Ouali, M. (2016). Gray wolf optimizer for hyperspectral band selection. *Applied Soft Computing*, *40*, 178–186. doi:10.1016/j.asoc.2015.09.045

Medjahed, S. A., Saadi, T. A., Benyettou, A., & Ouali, M. (2016). Gray wolf optimizer for hyperspectral band selection. *Applied Soft Computing*, *40*, 178–186. doi:10.1016/j.asoc.2015.09.045

Mirjalili, S. (2015). The ant lion optimizer. *Advances in Engineering Software*, *83*, 80–98. doi:10.1016/j.advengsoft.2015.01.010

Mirjalili, S., & Lewis, A. (2016). The whale optimization algorithm. *Advances in Engineering Software*, *95*, 51–67. doi:10.1016/j.advengsoft.2016.01.008

Mirjalili, S., Mirjalili, S. M., & Lewis, A. (2014). Grey wolf optimizer. *Advances in Engineering Software*, *69*, 46–61. doi:10.1016/j.advengsoft.2013.12.007

Mirjalili, S., Saremi, S., Mirjalili, S. M., & Coelho, L. D. S. (2016). Multi-objective grey wolf optimizer: A novel algorithm for multi-criterion optimization. *Expert Systems with Applications*, *47*, 106–119. doi:10.1016/j.eswa.2015.10.039

Panwar, L. K., Reddy, S., Verma, A., Panigrahi, B. K., & Kumar, R. (2018). Binary grey wolf optimizer for large scale unit commitment problem. *Swarm and Evolutionary Computation*, *38*, 251–266. doi:10.1016/j.swevo.2017.08.002

Sahoo, A., & Chandra, S. (2017). Multi-objective grey wolf optimizer for improved cervix lesion classification. *Applied Soft Computing*, *52*, 64–80. doi:10.1016/j.asoc.2016.12.022

Shakarami, M. R., & Davoudkhani, I. F. (2016). Wide-area power system stabilizer design based on grey wolf optimization algorithm considering the time delay. *Electric Power Systems Research*, *133*, 149–159. doi:10.1016/j.epsr.2015.12.019

Sharma, N., & Gedeon, T. (2014). Modeling a stress signal. *Applied Soft Computing*, *14*, 53–61. doi:10.1016/j.asoc.2013.09.019

Wang, L., Zheng, X. L., & Wang, S. Y. (2013). A novel binary fruit fly optimization algorithm for solving the multidimensional knapsack problem. *Knowledge-Based Systems*, *48*, 17–23. doi:10.1016/j.knosys.2013.04.003

Wang, M., Chen, H., Li, H., Cai, Z., Zhao, X., Tong, C., & Xu, X. (2017). Grey wolf optimization evolving kernel extreme learning machine: Application to bankruptcy prediction. *Engineering Applications of Artificial Intelligence*, *63*, 54–68. doi:10.1016/j.engappai.2017.05.003

Yang, X. S. (2010). Firefly algorithm, Levy flights and global optimization. In *Research and development in intelligent systems XXVI* (pp. 209–218). London: Springer. doi:10.1007/978-1-84882-983-1_15

Yang, X. S. (2011). Algorithm for multi-objective optimization. *International Journal of Bio-inspired Computation*, *3*(5), 267–274. doi:10.1504/IJBIC.2011.042259

Zhang, S., & Zhou, Y. (2017). Template matching using grey wolf optimizer with lateral inhibition. *Optik (Stuttgart)*, *130*, 1229–1243. doi:10.1016/j.ijleo.2016.11.173

KEY TERMS AND DEFINITIONS

Crossover: Means the recombination that means a genetic operator used to combine the genetic information of two parents to generate new offspring.

Energy Equation: Energy, in physics, is the capacity for doing work. It may be potential, kinetic, thermal, electrical, chemical, nuclear energy, etc.

Local Optima: In optimization, the local optima are the relative best solutions within a neighbor solution set.

Meta-Heuristic Algorithms: A metaheuristic is a higher-level procedure or heuristic designed to find a good solution to an optimization problem, especially with incomplete or imperfect information or limited computational capacity.

Multi Population: Multi population divides the whole population into multiple subpopulations. The population diversity can be maintained because different subpopulations can be located in different search spaces.

Optimization: Optimization means the process of finding the best solution(s) of a particular problem(s).

Swarm Intelligence (SI): It is basically population-based population. It depends on natural behavior of animals, bird etc. It utilizes the concept that a number of particles (candidate solution) which fly around the search space to find the best solution.

Chapter 3
Analog Integrated Circuit Optimization:
A 0.22µw Bulk–Driven Telescopic OTA Optimization Using PSO Program for Low Power Biomedical Devices

Houda Daoud
ENIS, University of Sfax, Tunisia

Dalila Laouej
ENIS, University of Sfax, Tunisia

Jihene Mallek
ENIS, University of Sfax, Tunisia

Mourad Loulou
ENIS, University of Sfax, Tunisia

ABSTRACT

This chapter presents a novel telescopic operational transconductance amplifier (OTA) using the bulk-driven MOS technique. This circuit is optimized for ultra-low power applications such as biomedical devices. The proposed the bulk-driven fully differential telescopic OTA with very low threshold voltages is designed under ±0.9V supply voltage. Thanks to the particle swarm optimization (PSO) algorithm, the circuit achieves high performances. The OTA simulation results present a DC gain of 63.6dB, a GBW of 2.8MHz, a phase margin (PM) of 55.8degrees and an input referred noise of 265.3nV/√Hz for a low bias current of 52nA.

DOI: 10.4018/978-1-7998-1718-5.ch003

INTRODUCTION

The tremendous growth in microelectronics over the past few decades have raised the CMOS technology miniaturization to a robust and an inexpensive electronic platform that is well suited for the implementation of wireless devices which they are essential parts of our everyday life. The wireless technology is used to meet the wireless devices requirements, therefore, we can move from person-to-person wireless communication towards person-to-device and device-to-device communication. Wireless technology has enabled us to obtain wearable wireless communication systems with higher speed, very small size and very low power consumption used for health monitoring system (Figure 1) (Mourad, 2018). In fact, the wearable and the implanted health monitoring technology has a strong potential to change the healthcare services future by enabling ubiquitous patients monitoring and it focuses on remote monitoring of elderly or chronically ill patients in residential environments.

Mobile health-monitoring devices offer great potential help for such patients who may be able to afford good healthcare without having to regularly visit their doctor. These technologies bring potential benefits to both patient and doctor. Doctors can focus more on priority tasks while saving time normally spent with consulting with patients who can move in their environment without having to make expensive trips when visiting their doctors, especially if they reside in remote locations. The mobile devices in both developed and developing countries present an opportunity to improve the health outcomes through the innovative delivery of health services and information. In the present domain of information technology, the wireless body area network has emerged as a new technology for e-healthcare that allows the data of a patient's vital body parameters and movements to be collected by small wearable or implantable sensors and communicated using short-range wired or wireless communication techniques (Mohammad, 2013). This has shown a great potential in improving the healthcare quality, and thus it has found a wide range of applications from ubiquitous health monitoring and computer assisted rehabilitation to emergency medical response systems. Through the health monitoring system, real-time and continuous triage information can be distributed to health care providers. Light weight and no-intrusive biomedical sensors like pulse oximeter and electrocardiogram are easily deployed for continuously monitoring of the vital patient signs and deliver the data to the first responders. Mobile healthcare systems are regarded as a solution to healthcare costs without reducing the patient care quality. Common architectures for health monitoring system involve; wireless sensor networks and smart phone technology have opened up new opportunities in the health monitoring system. The integration of the existing specialized medical technologies with cell phone and wireless sensor networks is a very promising application in home monitoring, medical care, emergency care and disaster response. However, the accessibility and the privacy protection of the data collected from a body area network, personal area network and wide area network is a major challenge. These transmitted data require a high demand for immediate access with security and practicality. The security and the privacy protection of the data collected through the Bluetooth enabling the mobile health monitoring system connected to the patient are another major challenges (Samaher, 2017), (Abayomi, 2014). Today, the researchers are focused on biomedical field while the size and the power consumption are the main critical issues for the implantable devices. To attain these challenges, many techniques and methods are investigated. Indeed, through the improvements and the miniaturization in MOSFET technologies we can obtain medical dispositive with very small size while the power consumption is a few hundreds of micro watts. Figure 2 presents the wireless micro system structure which it consists of four modules: the acquisition unit, the processing unit, the power unit and the transmission unit. The RF transmitter architecture is given by Figure 3. A biomedical device is a combination of two parts: hard and soft. For the hard part, the transmission unit is the most greedy in terms of power consumption (Table 1) (Andrianiaina, 2015). Several processing techniques can be used for such purposes time or frequency domain methods including filtering, averaging, spectral estimation, and others. Even if it is possible to deal with continuous time waveforms, it is usually convenient to convert them into a numerical form before processing. The recent progress of digital technology, in terms of both hardware and software, makes digital rather than analog processing with good efficiency and flexibility. Digital techniques have several advantages: their performance is generally powerful, being able to easily implement even complex algorithms, and accuracy depends only on the truncation and round-off errors, whose effects can be predicted and controlled by the designer and they are largely unaffected by other unpredictable variables such as component aging and temperature, which can degrade the performances of analog devices. Moreover, design parameters can be more easily changed because they involve software rather than hardware modifications (Luca, 2006).

An implant device must satisfy several constraints where the most critical ones are the size the power consumption and the lifetime. The analog-to-digital conversion is a vital step to achieve the communication between the two digital and analog parts. Among the diversity of operational amplifiers architectures, the operational transconductance amplifier (OTA) is the most promising candidate to meet the high performances analog-to-digital convertors (ADC) needs. This circuit is a very critical block in the design of a variety of analogue integrated circuits such as current conveyor (CCII), filters, integrators and so on (Raj, 2013). For the circuit implementation, switched capacitors (SC) circuits are still good candidates for very low-voltage wearable wireless communication systems (Dima, 2016).

For a biomedical dispositive, the OTA must be polarized by a very low voltage and it has to consume as low as possible power. Generally, a gate-driven NMOS to be

Figure 1. General overview of the remote health monitoring system

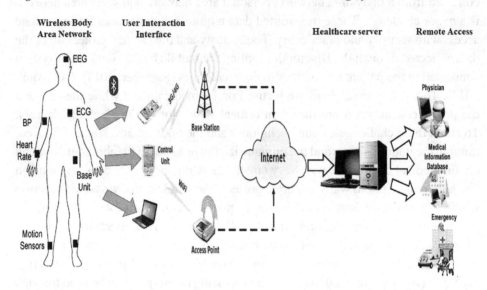

Figure 2. Wireless microsystem structure

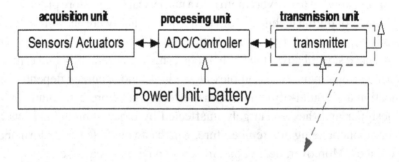

Figure 3. RF transmitter architecture

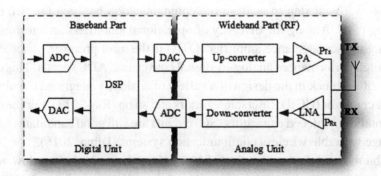

Table 1. Portion of power consumed by different modules

Modules	Power Consumption Portion (%)
Measurement and instrumentation	2
Micro-controller	25
Transmitter	73

conductor, the gate-source voltage V_{gs} must be larger than the threshold voltage V_{th}. Therefore, the last voltage limits the implementation of the analogue circuit at low voltage. In fact, for very low power applications, the MOSFET must be polarized in weak inversion (Matej, 2017). To achieve this objective, the Bulk-Driven technique is a good solution since it permits to overcome the threshold voltage limitations while we can increase or decrease the MOS transistor threshold voltage (Matej, 2017). Consequently, the Bulk-Driven technique allows the operation in the moderate inversion region at a supply voltage equal to the threshold voltage V_{th} of the used technology. The Bulk- Driven method principle is that; the gate source voltage (V_{gs}) is set to a sufficient value to form an inversion layer while the input signal is applied to the bulk terminal (Viera, 2018). According to the litterature, this technique was used in many OTA structures such as the one stage folded cascode OTA, the two-stage OTA and the folded cascode OTA (Arash, 2012), (Hassan, 2010).

In this chapter, the authors are focused on low power Bulk-Driven Telescopic OTA optimization working for frequencies that lead to a biomedical device design for ultra-low power applications. The main objective of section 2 is to highlight the OTA circuit role and utility in different applications. After having exposed the different OTA architectures specificities, we will opt for one structure which is dedicated for low power applications. The chosen OTA architecture will be more improved to meet the requirement of an ultra low power biomedical application. This chapter presents the Bulk-Driven technique principle in section 3. Section 4 describes the proposed Telescopic OTA architecture. Analytical design equations defining the Bulk-Driven Telescopic OTA performances are investigated. Section 5 presents an algorithmic driven methodology for the proposed OTA optimization where the Particle Swarm Optimization (PSO) approach is well explained. In this section, this chapter presents the optimized OTA performances, discusses the obtained simulated results, besides, a comparison between the optimized Bulk-Driven telescopic OTA performances and some published works are introduced. In section 6, the authors apply the Bulk-Driven OTA circuit in a SC integrator. Finally, section 7 gives some concluding remarks drawn from this chapter and presents the scope for future research.

OPERATIONAL TRANSCONDUCTANCE AMPLIFIER (OTA)

The OTA is a fundamental block of the CAN which, using the switched capacitor technique, forms the most critical integrator of the entire circuit. The CMOS technology makes it possible to combine digital functions and analog functions on the same substrate. Having good performances is achieved at the cost of a greater design complexity that it can only be attained through the use of high efficiency simulation tools. Designing analog circuits that perform so well in the base band is becoming increasingly challenging with the persistent trend towards lower power supply voltages. In order to improve certain characteristics (high gain, low bias current, low dissymmetry,.. etc.), manufacturers are constantly developing various integrated amplifier structures that are often very complex. As a result, the circuit design still appears today as a complex activity that requires a well-studied design methodology. Currently, operational amplifiers (OAs) are widely used as building blocks implemented in analog applications. OAs perform well for low frequency applications such as audio and video systems (Walt, 2005). However, for high frequencies, the design of OAs becomes difficult due to their frequencies limitations. As a result, the OTAs replace the OAs to become a fundamental block for high and low frequency applications (Nikhil, 2010). With respect to the efforts devoted by integrated circuit researchers and to the continued progress of commercial semiconductor technology, the OTAs can operate for a variety of frequencies.

Description

An OTA is defined as an amplifier with low impedance nodes except the input and the output nodes, this gives it the possibility to work at very high frequencies. It is essentially a voltage-current amplifier in which the output current corresponds to the input voltage multiplied by the transconductance. Figure 4 illustrates the symbol of a transconductance operational amplifier.

The OTA has two terminals for the differential input, a current output I_{out} and a transconductance g_m which can be adjusted by a bias current I_{bias}. The OTA charge is generally capacitive. Its output resistance is high.

OTA History and Application Fields

In the early 1990, bipolar and GaAs technologies dominate the Radio Frequency Integrated Circuits (RFIC), while the CMOS technologies are used for baseband signal applications (Ahmadreza, 1996). As soon as the gate length of the CMOS transistors is decreased to 1 µm in the mid-1990, the CMOS RFIC can reach gigahertz of frequencies. The submicron CMOS technology development creates a new track for

Figure 4. The OTA internal block diagram

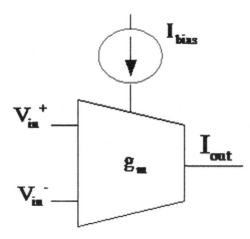

the CMOS RFIC design constituting the OTA circuits. The CMOS OTA has become an essential component in a number of electronic circuits in various applications. Figure 5 shows different applications that use frequencies ranging from few tens of hertz to several gigahertz (Edgar, 2004).

Among these applications, various are the medical electronics and the seismic sector where the frequency range is between 0.1Hz up to 20Hz. For the some MHz frequency range as in the intermediate frequency (IF) filters in RF receivers, the realizations of the Gm-C filters are very relevant. For applications in a range of some GHz, where the OTA is a fundamental block, a number of researchers are focused on filters and oscillators design. As the OTA circuit application field is large, the OTAs structures are different.

Figure 5. Application Fields

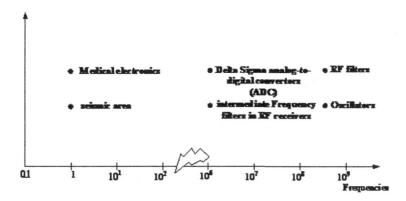

OTA Characteristics

Different characteristics exist to determine the OTA structure performances given in Figure 6. The dynamic and the static characteristics presented in Table 2 are distinguished.

The dynamic characteristics are the open-loop gain, the gain-bandwidth, the slew rate, the CMRR, the PSRR and the noise (thermal and flicker). Indeed, a frequency open-loop OTA circuit study makes it possible to determine the static gain, the gain-bandwidth product and the phase margin. The gain product must be high to meet the needs of the new telecommunication systems. The phase margin must be at least 45 degrees for the circuit stability circuit. Then, the common mode rejection ratio (CMRR) determination makes it possible to describe the OTA sensitivity circuit to the voltages at positive and negative inputs, and therefore the sensitivity of the OTA to reject a common signal. This feature is important in analog applications where signals are transmitted in differential modes. Then, to see the OTA sensitivity to the supply voltages variations V_{dd} and V_{ss}, the power noise rejection ratio (PSRR) is defined. In addition, low power applications require a small slew rate with low bandwidth. In addition, noise, which is the sum of the thermal noise and the flicker noise ($1/f$), must be minimized. From Table 2, the static characteristics are the output and the input dynamics, the output dynamic range, the effective transconductance, the supply voltage, the bias current and the power consumption. Indeed, an OTA follower configuration makes it possible to determine the dynamic output range and the dynamic input range which respectively represent the maximum positive or negative output voltage obtained for a zero static voltage at the output and the common mode voltage that can be applied to the OTA inputs without damaging its performance. The OTA circuit must have a wide signal dynamics to increase its output dynamics. In addition, the effective transconductance is a key parameter for determining the dynamic characteristics majority. The output dynamic range (DR) defines the ratio between the maximum signal power that the system can tolerate

Figure 6. Block diagram of the OTA circuit

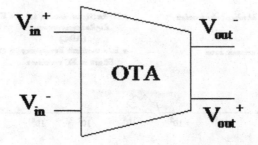

without signal distortion and the system noise level. More the noise is minimized and the signal dynamic is expanded, more the output dynamics is increased. Wireless communication systems require OTA circuits with low power consumption where the supply voltage and the bias current are responsible for minimizing the circuit power consumption. The dynamic and the static characteristics will be taken into consideration when designing such an OTA structure that represents a fundamental block used in the new mobile systems.

Diversity of OTA Topologies

The OTAs topologies are very varied for a very large and specific field of application. Our research focuses on differential input and output structures because they exhibit a large output dynamic, eliminate the distortion problem, and reject noise from the substrate. According to the literature, several OTA structures exist, the best known

Table 2. Dynamic and static characteristics of the OTA circuit

Dynamic Characteristics		Static Characteristics	
Open loop gain (dB)	$A_V = \dfrac{V_{out}}{V_{in}}$	Output dynamic (V)	$[V_{out,min}; V_{out,max}]$
Gain-bandwidth product	$GBW = A_V f_{dp}$	Input dynamic (V)	$[V_{in,min}; V_{in,max}]$
Common mode rejection ratio (dB)	$CMRR = \dfrac{A_{Vd}}{A_{Vc}}$	Effective transconductance (S)	$g_{m,eff} = \dfrac{I_{out}}{V_{out}^+ - V_{in}^+}$
Power supply Rejection Ratio (dB)	$PSRR^+ = \dfrac{A_V}{V_{out/V^+}}$	Output dynamic range (dB)	$DR = \dfrac{P_{s,max}}{P_{noise}}$
Slew Rate (V/µs)	$SR = \dfrac{I_{bias}}{2C_{out}}$	Voltage supply (V)	V_{dd}
Thermal noise (V/√Hz)	$\overline{i_{th}^2} = 4KTnyg_m$	Bias current (mA)	I_{bias}
Flicker noise (V/√Hz)	$i_{1/f}^2 = \dfrac{KF}{C_{ox}WLF} g_m^2$	Power consumption (mW)	$P=(V_{dd}-V_{ss}).\sum I_{bias}$

of which are the following: the simple OTA, the two-stages OTA, the super AB class OTA, the folded cascaded OTA and the telescopic OTA (Arash, 2012), (Hassan, 2010), (Sombat, 2010), (Seyed, 2017), (Bellamkonda, 2016).

Basic Configuration OTA

Figure 7 presents the basic configuration of the OTA. The analytical expressions of the static gain and the gain-bandwidth product are given by the following relationships:

$$A_v = g_{m1} R_L = g_{m1} (r_{01} // r_{03}) \tag{1}$$

Figure 7. basic configuration OTA

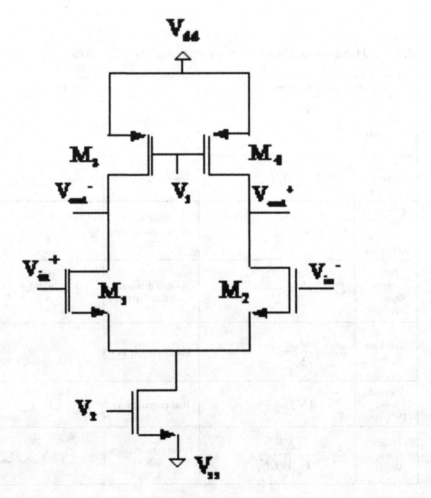

$$GBW = \frac{g_{m1}}{C_{out}} \qquad (2)$$

Where R_L, g_{m1} and C_{out} are respectively the output resistance, the transconductance of the M_1 transistor and the capacity at the output node. The resistance R_L is equal to $r_{01}//r_{03}$. The output resistance of the basic configuration OTA is relatively low, which limits its static gain and increases its transition frequency as well as its speed. To increase the output resistance, the two-stage OTA circuit is highlighted.

Two-stages OTA

One stage is added in Figure 8, hence the generation of the two-stages OTA. The two-stages OTA is often called the OTA Miller.

The static gain and the gain-bandwidth product are presented by the following expressions:

$$A_v = g_{m2}g_{m6} \, (r_{02}//r_{04})(r_{06}//r_{08}) \qquad (3)$$

$$GBW = \frac{g_{m2}}{C_C} \qquad (4)$$

Where g_{m2}, g_{m6} and C_C are respectively the transconductances of M_2, M_6 transistors and the compensation capacity. r_{02}, r_{04}, r_{06} and r_{08} are respectively the drain-source resistances of M_2, M_4, M_6 and M_8 transistors. Compared to the basic configuration OTA, the addition of the stage amplifies the gain but increases the complexity of the circuit which reduces its speed. In addition, the compensation elements constituting the R_C resistance and the C_C capacitance must be well designed to not lead to the stability problem which limits the transition frequency of the circuit.

Super Class AB OTA

In order to minimize the power consumption, the two-stages OTA circuit of Figure 8 is converted into a super-class AB OTA circuit (Michiel, 2008), (Jaime, 2002). Figure 9 shows a conventional OTA (Figure 9 (a)) that can be converted to a super-AB OTA (Figure 9 (b)). An adaptive bias circuit replaces the bias current I_{bias} of Figure 9 (a) and provides a very small current for the M_1 and the M_2 transistors. As soon as this circuit detects a wide differential input, it automatically increases the bias current supplied in the differential stage. The additional current increase is

Figure 8. Two-stages OTA

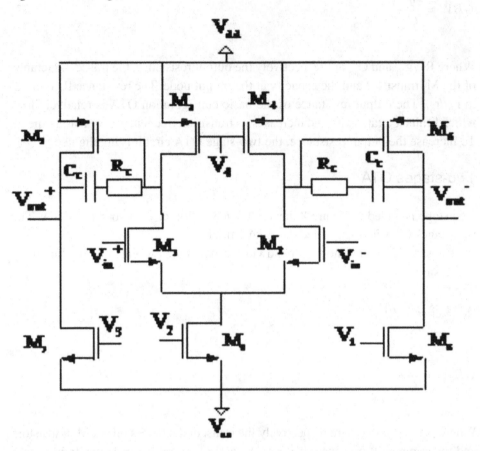

obtained by the local common mode feedback (LCMFB) illustrated by the R_1 and R_2 resistances then the gain-bandwidth product becomes high (Juan, 2007). The gain-bandwidth product of the super-class OTA AB in Figure 9 (b) is presented by the following expression:

$$GBW = \frac{g_{m1}g_{m7}R_X}{2\pi C_L} \tag{5}$$

The Figure 10 structure has been improved compared to the basic topology presented in Figure 9. Indeed, the gain is increased by using a second stage constituting the cascode transistors formed by the M_9 and the M_{11} transistors. Then, the R_1 and R_2 active resistances are controlled by M_{17} and M_{18} MOS transistors operating in ohmic regime. These transistors are controlled by the V_c voltage which will be chosen to satisfy the phase margin and stabilize the circuit.

The adaptive circuit is replaced by the "Filpped Voltage Follower" (FVF) circuit formed by M_3 and M_5 transistors. This circuit provides adaptive bias: low current for zero differential input and high current proportional to the dynamics of the differential input signal. For switched capacitor circuits, the OTA bias current will be regulated to provide a strong current at the beginning of the charge transfer phase and then increase the circuit speed, after which the current begins to decrease gradually when the charge will be transferred which increases the circuit gain. Although the power consumption is reduced, the OTA class AB structure remains limited in terms of frequency when used in broadband applications that require hundreds of megahertz of frequencies (Mohammad, 2005).

Folded Cascode OTA

In the literature, several folded cascode structures are presented and used in low power applications (Arash, 2012). Figure 11 shows the folded cascode OTA which is a typical architecture for high frequency applications. This circuit contains a differential stage formed by M_9 and M_{10} transistors. The input stage differential pair can be implemented either by NMOS transistors (Figure 11 (a)) or by PMOS transistors (Figure 11 (b)). Indeed, the PMOS input stage has a lower transconductance than the NMOS type. This is due, essentially, to the high mobility of the NMOS transistors. In addition, the PMOS transistor provides a lower flicker noise. The

Figure 9. (a) Class A OTA, (b) Super class AB OTA

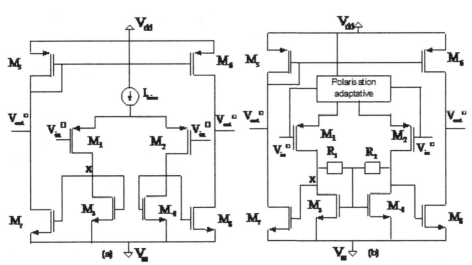

Figure 10. Improved super class AB OTA

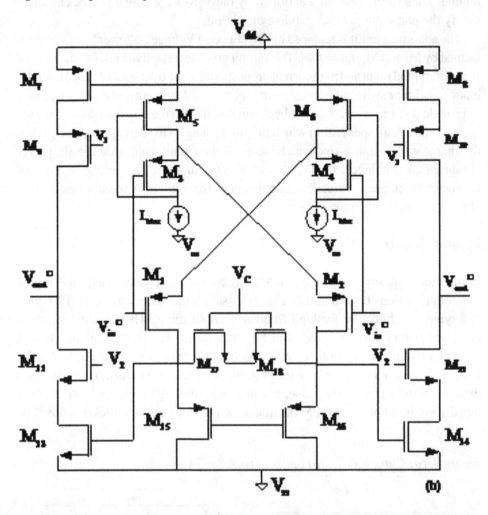

(b)

static gain and the gain-bandwidth product of Figure 11 (b) are presented by the following expressions:

$$A_V = g_{m9}\left(g_{m3}r_{03}r_{01}\right) // \left(g_{m5}r_{05}\left(r_{07} // r_{09}\right)\right) \tag{6}$$

$$GBW = \frac{g_{m9}}{C_{out}} \tag{7}$$

Where g_{mi} is the transconductance of the M_i transistor with i = (3, 5, 9). r_{0i} is the drain-source resistance of the transistor M_i with i = (1, 3, 5, 7, 9). C_{out} is the OTA circuit output capacitance.

Telescopic OTA

Telescopic structures are widely used due to their low noise and power consumption [20]. Figure 12 shows the telescopic OTA structure. This circuit constitutes a differential pair formed by the M_1 and M_2 transistors. The M_3 and M_5 cascode transistors play the role of the output impedance increasing, then the static gain is high. The V_1, V_2, V_3 voltages are the bias voltages. The I_{bias} bias current feeds the differential stage. The V_1 and the V_2 voltages can be replaced either by the Wilson mirror or by the Cascode mirror because of their high output resistances and either by simple current mirrors. The static gain and the gain-bandwidth product are given by the following expressions:

$$A_V = g_{m1}\left(g_{m5}r_{05}r_{07}\right)//\left(g_{m3}r_{03}r_{01}\right) \tag{8}$$

Figure 11. (a) Folded cascode OTA with NMOS input, (b) Folded cascode OTA with PMOS input

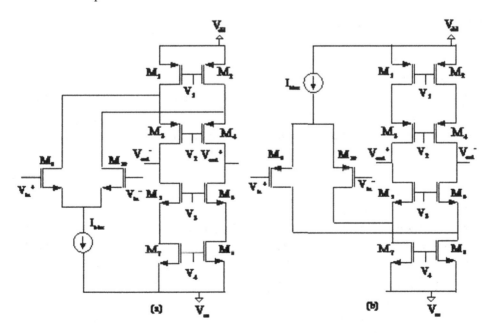

$$GBW = \frac{g_{m1}}{C_{out}} \quad\quad\quad (9)$$

Where g_{m1}, g_{m3} and g_{m5} are the transconductances of M_1, M_3 and M_5 transistors. r_{01}, r_{03}, r_{05} and r_{07} are respectively the drain-source resistances of M_1, M_3, M_5 and M_7 transistors. C_{out} is the capacitance at the output of the telescopic OTA circuit.

The OTA Telescopic remains the best candidate for low power biomedical device design due to its low noise and power consumption. Therefore, in this chapter, the authors are looking for enhancing the telescopic performances in terms of power consumption by using the Bulk-Driven technique.

BULK-DRIVEN TECHNIQUE

In the last few decades, the analog designers devote recently their efforts to low-voltage low-power integrated circuits design since the power consumption has become a critical issue. In fact, integrated circuits design has recently aimed to low-voltage and low-power applications, especially in the wearable wireless communication system environment where a low supply voltage, given even by a transmitter, is used. During nanoscale process, the future CMOS device threshold voltage is not expected to decrease much below. Hence a good solution to overcome the threshold voltage limitation is to use the Bulk-Driven technique.

Bulk-Driven Principle

In the conventional operational amplifier design, the input signal is usually applied to the gate terminal while the bulk terminal is tied to V_{dd} or V_{ss} voltages, as shown in Figure 13. (a), where the drain current is controlled by the V_{gs} voltage. Contrary, when using the Bulk-Driven technique (Figure 13. (b)), the input signal is injected into the input transistor bulk terminal while the gate is biased by a fixed voltage ($V_{gs} > V_{th}$ for NMOS) to form an inversion layer (Kshitij, 2016).

Bulk-Driven Operation

For a Gate-Driven NMOS, the transistor to be conductive, the V_{gs} voltage must be larger than the V_{th} one as it is indicated in Figure 14 (in this case, it is approximately 0.48V). According to equation 10, the use of this technique decreases the threshold voltage value and the transistor is active even if the input is negative, zero or positive.

Figure 12. Telescopic OTA

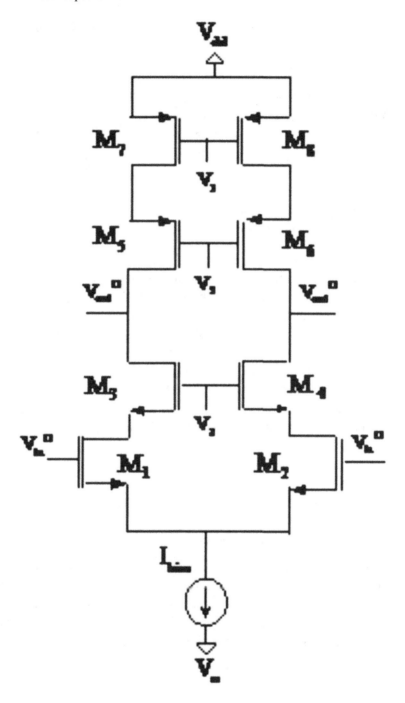

Figure 13. (a) Conventional Gate-Driven, (b) Bulk-Driven NMOS

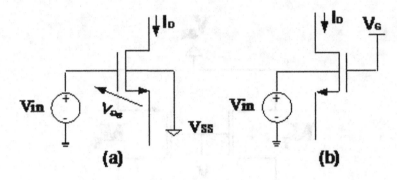

(a)   (b)

$$V_{th} = V_{T0} + \gamma \left(\sqrt{2|\varphi_F| - V_{bs}} - \sqrt{2|\varphi_F|} \right) \tag{10}$$

Where, V_{T0}, γ and φ_F are respectively the zero bias threshold voltage, the body effect coefficient and the Fermi potential. Using the Bulk-Driven technique allows to decrease the threshold voltage. Therefore, this technique has an important advantage such it allows to operating at low supply voltage while the input voltage is negative, zero and even few positive used for biasing the transistor to attain the desired drain currents (Nikhil, 2010) (Figure 14). From this curve, in the saturation region, the V_{ds} voltage has no influence on the I_D current. Thus, this current is only influenced by the V_{bs} voltage. In fact, while the V_{bs} increases, the I_D increases. Accordingly to equation (11), this current increases when the V_{gs}, the V_{bs} or the W/L increase.

$$I_D = \frac{\beta}{2} \left[V_{gs} - V_{T0} + \gamma \left(\sqrt{2|\varphi_F| - V_{bs}} - \sqrt{2|\varphi_F|} \right) \right]^2 (1 + \lambda V_{ds}) \tag{11}$$

Where $\beta = \mu C_{ox} \dfrac{W}{L}$ and $\lambda = 1/V_A$, where V_A is the Early voltage.

Since this technique has the advantage of the operation possibility in the weak inversion region, it has some drawbacks such as the g_{mb} body transconductance which it is approximately five times smaller than the g_m gate transconductance (Kshitij, 2016). It causes, consequently, lower value of OTA unity gain frequency. In addition, this structure presents large parasitic capacitance on the bulk and high input referred noise.

Figure 14. I-V Bulk-Driven MOSFET Characteristic for $V_{gs}=0.5V$

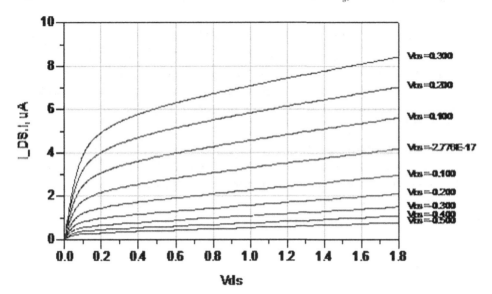

THE BULK-DRIVEN TELESCOPIC OTA ARCHITECTURE

the Bulk-Driven technique is used briefly in this section to build some of the basic analog building blocks, especially the Telescopic OTA.

Proposed Bulk-Driven Telescopic OTA Architecture

An operational transconductance amplifier is a voltage-current amplifier when the output current is proportional to the input voltage by the transconductance. It is a critical block that it can be designed for realizing a variety of integrated circuits. The designer must opt for the appropriate architecture to alleviate the trade-off between gain, speed, linearity and power consumption. In fact, the telescopic OTA is a good candidate for low power consumption applications. Indeed, this structure is low noisy and it is the fastest compared with other structures, but, it is limited in terms of output swing. In this work, the authors choose to apply the Bulk-Driven technique to a telescopic OTA in order to reduce the power consumption. The proposed OTA structure is presented in Figure 15. In this circuit, the inputs are applied in the differential input stage (M_1- M_2) transistors bulk where the g_{mh} body transconductance of the input transistors provides the input transconductance. The differential input stage is polarized by a simple current mirror (M_9, M_{10}).

Circuit Analysis

Since the medical dispositive design should be implemented using analog integrated circuits with very small size, this subsection presents all the performances of the proposed OTA circuit which are closely related to the transistors scaling. An operational transconductance amplifier is characterized by many features such as; DC gain, Gain Bandwidth (GBW), Slew Rate (SR), CMRR, PSRR, input referred noise and so on. The application of the input signal in the bulk of the transistor will change all the OTA performances expressions which they depend on the bulk. Then, a small signal analysis of the Bulk-Driven telescopic OTA is carried out to explicit the different characteristics intended optimized. From the proposed telescopic OTA structure presented in Figure 15, the open-loop voltage gain and gain bandwidth are given by the following expressions:

$$A_V = g_{mb1}.\left(R_{out1} // R_{out2}\right) \tag{12}$$

Where

$$R_{out1} = g_{m5}.r_{05}.r_{07} \text{ and } R_{out2} = \frac{-1 - \dfrac{g_{ds1}}{g_{ds3} + g_{m3}} + \dfrac{g_{m3}}{g_{ds3} + g_{m3}}}{\dfrac{g_{ds3}.g_{m3}}{g_{ds3} + g_{m3}} - \dfrac{g_{ds1}.g_{ds3}}{g_{ds3} + g_{m3}} - g_{ds3}} \tag{13}$$

g_{m3} and g_{m5} are respectively the transconductances of the NMOS transistor M_3 and the PMOS transistor M_5. g_{ds1} and g_{ds3} are respectively the conductances of M_1 and M_3 transistors. g_{mb1} is the bulk transconductance of the NMOS transistor M_1.

$$GBW = \frac{g_{mb1}}{2\pi C_L} \tag{14}$$

The positive power-supply rejection ratio (PSRR) is expressed as:

$$PSRR = \frac{((R_{01} + R_{02} + r_{03}).(r_{03} + R_{01} + g_{m3}.r_{03}.R_{01}))}{R_{02}.((1 + g_{m3}.r_{03}))} \tag{15}$$

Where

Figure 15. Proposed Bulk-Driven Telescopic OTA

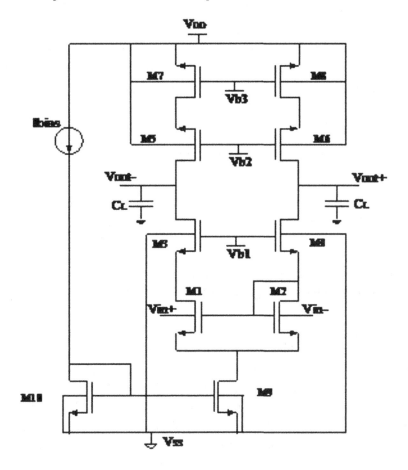

$$R_{01} = \frac{r_{01}r_{09}}{(r_{01} + r_{09})} \; and \; R_{02} = r_{05}r_{07}g_{m5} \tag{16}$$

Where r_{01}, r_{03}, r_{05}, r_{07} and r_{09} are the drain-source resistances of transistors M_1, M_3, M_5, M_7 and M_9. The common mode rejection ratio (CMRR) can be approximated as:

$$CMRR = \frac{A_v}{A_c} \; where \; A_c = \frac{r_{07}}{2r_{09}} \tag{17}$$

The input referred thermal noise voltage of the Bulk-Driven telescopic OTA can be presented as:

$$\overline{V_{in,th}^{2}} = 4kT\left(2\frac{2}{3g_{mb1}\cdot g_{m2}} + 2\frac{2g_{m2,4}}{3g_{mb1}\cdot g_{m2}}\right) \tag{18}$$

Where k is the Boltzmann's constant and T is the temperature. The input referred flicker noise voltage of the used OTA can be written as:

$$\overline{V_{in,1/f}^{2}} = 2\frac{KFN}{C_{ox}\left(WL\right)_{1,2}\cdot f} + 2\frac{KFP}{C_{ox}\left(WL\right)_{7,8}\cdot f}\frac{g_{m7,8}^{2}}{3g_{mb1}\cdot g_{m2}} \tag{19}$$

Where KF is the flicker noise coefficient, f is the frequency; C_{ox} is a constant for a given process. W and L are the sizes of transistors.

THE BULK-DRIVEN TELESCOPIC OTA DESIGN WITH PSO ALGORITHM

Expert analogue designers devote their efforts to make a performant analogue integrated circuit design since it is still a hard task. Being the complex relationship between process parameters, transistor sizes and design requirements, analog integrated circuits automatic design is a big challenge. An efficient search in the optimization tools is mandatory when hard specifications of analogue building blocks must be satisfied, mainly for low voltage and low power applications. According to the literature, several tools automating the topology synthesis, were proposed. A synthesis procedure based on the g_m/I_D methodology is introduced by Flandre and Silveira (Flandre, 1996). This method using the universal g_m/I_D as a function of $I_D/$ (W/L) characteristic of the CMOS technology under consideration, clarifies a top-down synthesis methodology for CMOS analogue circuits architectures. In order to further improve the CMOS circuits performances by identifying the parameters keys acting on circuit performances to lead to an optimal design, optimization process which it represents a group of individual optimization runs was developed (Manik, 2016). In fact, (William, 1988) present sizing of nominal circuits. Sizing problems accounting for parameter tolerances and worst-case optimization were addressed. In addition, optimization tool based on heuristic algorithms was investigated in many analog circuit designs and gave promising results (Hasançebi, 2012). Optimization method based on PSO algorithm was introduced in (Wei, 2018). It is an evolutionary computation method based on the social behavior and the movement of swarm searching for the optimal and the best location. PSO is a stochastic global optimization method which is based on simulation of fish and birds social behavior.

As in Evolutionary Algorithms such as the Genetic Algorithm, PSO exploits a population of potential solutions to probe the search space. In this algorithm, each single solution is a particle in the search space. All of the particles or solutions have fitness values which are evaluated by the fitness function to be optimized, and they have velocities which direct the flying of the particles. The particles fly through the problem space by following the current optimum solutions (Sameh, 2010).

Problem Formulation

The telescopic OTA has many features such, DC gain, gain bandwidth product, CMRR, PSRR, SR and so on. This circuit is based on a number of CMOS transistors. Thus, this features measures are determined by the design parameters. Consequently, in our work, this algorithm, which is implemented in MATLAB, is used in order to improve the Bulk-Driven Telescopic OTA performances. Since, this algorithm was used for maximizing or minimizing an objective function. In this term, the main objective is to size the width and the length of the CMOS transistors and the appropriate current. The OTA specifications are the same as given by Table 1. We developed an objective function which increases the static gain A_v, the transition frequency f_T, the PSRR and the CMRR. Besides, the objective function aims to decrease the input referred noise and the area of the OTA circuit. In this term, our main purpose is to size the width and the length of the CMOS transistors and the appropriate current while the objective function is presented by Eq.11.

$$Fo = a_1.A_v + a_2.f_t + a_3.CMRR + a_4.PSRR + a_5.\frac{1}{V_{T,in}^{2}} + a_6.\frac{1}{\sum W_i L_i} \tag{20}$$

Where a_{1-6} are the coefficients used to normalize the different sizes.

PSO Algorithm Approach

PSO is initialized with a group of random particles and then look for optima by updating generations. In every iteration, each particle is updated by the following two "best" values. The first one is the best solution (fitness) it has achieved so far. The fitness value is also saved. This value is called pbest. Another "best" value that is tracked by the particle swarm optimizer is the best value, obtained so far by any particle in the population. This best value is a global best and it is called gbest. When a particle takes part of the population as its topological neighbors, the best value is a local best and it is called lbest. After finding the two best values, the particle updates its velocity and positions with the following equations (21) and (22).

$$v[\ ] = v[\ ] + c1.rand(\ ).(pbest[\ ] - present[\ ]) + c2.rand(\ ).(gbest[\ ] - present[\ ])$$

(21)

Where,

$$present[\ ] = present[\ ] + v[\ ]$$

(22)

v[ ] is the particle velocity, present[ ] is the current particle (solution). pbest[ ] and gbest[ ] are defined as stated before. rand () is a random number between (0, 1). c_1 and c_2 are the learning factors. usually, $c_1 = c_2 = 2$. The flow chart of the PSO algorithm for designing a Bulk-Driven Telescopic OTA is presented in Figure 16.

Simulation Results

To evaluate the proposed OTA structure, this chapter presents the simulated OTA performances. For this type of circuit, the DC gain is one of the most important parameters. It indicates the capacitance of circuit to amplify the inputs signals. According to Figure 17, the proposed OTA has an important DC gain of 64.6dB and a GBW of 1.4MHz. To prove the stability of the circuit, the phase marge (PM) must be more than 45 degrees. Therefore, the optimized OTA is stable with a PM of around 65 degrees (Figure 17). According to Figure 18, Figure 19, Figure 20, the optimized telescopic OTA presents a CMRR of 54.4dB and a PSRR of 36.7dB, it has a very low input referred noise. Table 3 resumes the optimized Telescopic OTA Performances.

OTA Performances Comparison With Published Works

The proposed bulk-driven telescopic OTA is implemented using TSMC 0.18μm CMOS technology. Therefore, this section presents a summary of the OTA performances that are compared to other recent published works. From Table 4, compared to the published works, the Bulk-Driven OTA structure can offer much better performance in terms of static gain and power consumption. Indeed, the optimized proposed OTA has a gain and a power consumption respectively of the order of 64.6dB and 0.22μW.

Figure 16. Flow chart of PSO

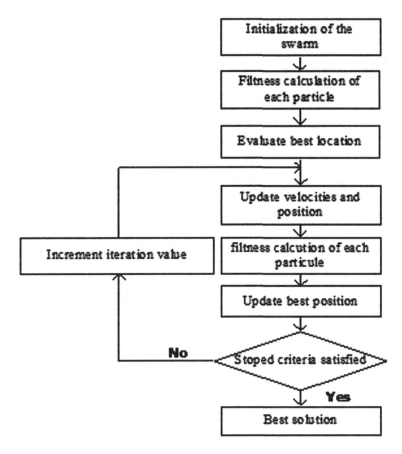

SWITCHED CAPACITOR INTEGRATOR APPLICATION

The switched capacitance technique has long dominated the analogue circuit integrated design field. It is a mature technology providing a number of commercial circuits. The first switched capacitor circuit dates since 1980. These circuits use a smaller silicon surface and they are generally less expensive. The principle of the switched capacitance technique constitutes an alternative approach of circuits design allowing to realize various functions such as filters, oscillators, analogue to digital convertors, ...ect. For implementation, the switched capacitor circuits remain the best candidates for low power applications. In switched capacitors circuits, the OTA which is configured as an integrator is the principle building block for SC filters and ADC. The proposed Bulk-Driven Telescopic OTA satisfies the gain and the speed requirement of SC circuits applications.

Figure 17. Gain and phase marge curve

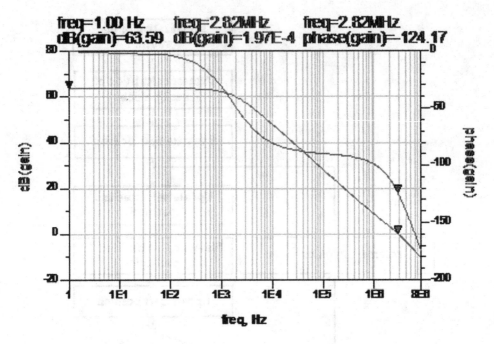

Figure 18. CMRR curve

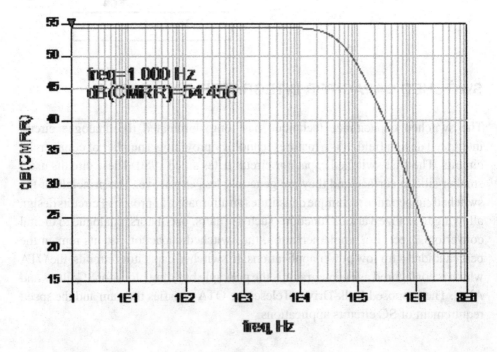

Figure 19. PSRR curve

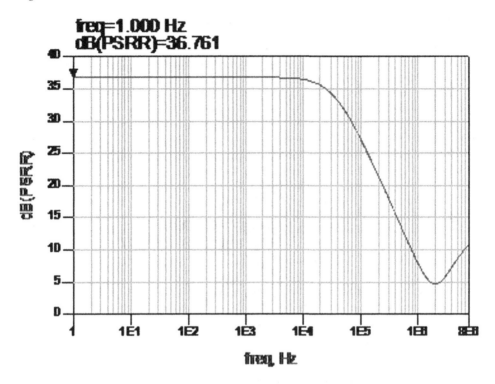

To underline the suitability of the proposed simulated OTA, it is applied in a first-order SC integrator (Figure 21). The output of the SC integrator can be given by the following expression:

$$V_{out}(z) = \frac{C_s}{C_f} \frac{z^{-1}}{1-z^{-1}} V_{in}(z) \tag{23}$$

The integrator is controlled by two-phases with non-overlapping clocks Φ_1 and Φ_2, where the input signal is sampled on Φ_1 and integrated on Φ_2. Figure 22 illustrates the required non-overlapping clock signals.

During the integration phase, the charge is transferred from the C_S sampling capacitor to the C_f feedback capacitor and the op amp output must settle to the desired level of accuracy within the selected time. In addition, the SC integrator is realized using MOS switches and capacitors. Since the optimized Bulk-Driven Telescopic OTA circuit is ready to use, the switches and the C_S and the C_f capacitors design are necessary.

Figure 20. Input referred noise curve

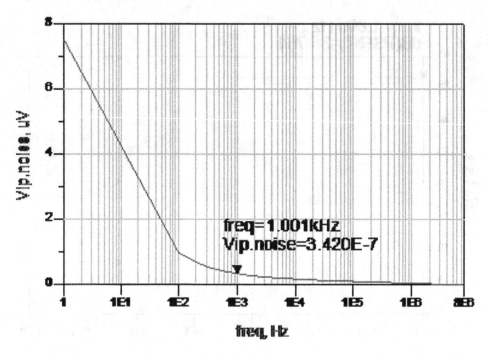

Table 3. Optimized Telescopic OTA Performances

Performances (C_L=0,1 pF)	Simulations Results
DC Gain (dB)	63.6
GBW (MHz)	2.8
Phase Margin	55.8
CMRR(dB)	54.4
PSRR(dB)	36.7
SR (V/µs)	±0.09
Offset voltage (mV)	0.6
Output swing (mV)	[-850, 556]
Input referred noise voltage at f=1 kHz (µV/$\sqrt{\text{Hz}}$)	0.265
Bias current (nA)	52
Pc (nW)	223

Table 4. OTA performances summary and comparison with published works

References	(Elena, 2016)	(Kumar, 2017)	(Jonathan, 2003)	(Karima, 2018)	This Work
CMOS Technology (µm)	0.18	0.18	0.18	0.18	**0.18**
Supply Voltage (V)	0.5	±0.6	0.7	0.4	**±0.9**
DC gain (dB)	58.4	54	57.5	60	**63.6**
GBW (MHz)	18	4	3	0.347	**2.8**
CMRR (dB)	87.9	144	19	82.38	**54.45**
PSRR (dB)	94.3	158	52.1	-	**36.76**
Phase margin	-	63	60	57	**55.8**
Bias current (µA)	0.1	-	36.3	-	**0.052**
DC offset voltage (mV)	-	-	11	-	**0.6**
Input referred noise voltage (µV/√Hz)	-	0.8	0.1	-	**0.265**
Transconductance (µS)	-	5.3	-	96.12	**0.8**
C_L(pf)	1	1	20	-	**0.01**
P_C (µW)	0.62	40.14	25.4	0.28	**0.223**

Figure 21. SC integrator using the optimized Bulk-Driven Telescopic OTA

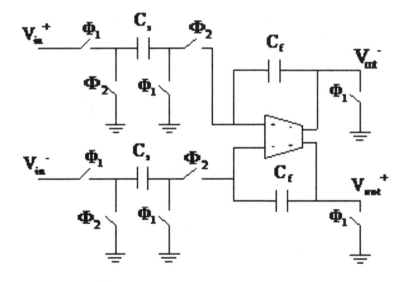

Figure 22. Clock signals used in the SC integrator

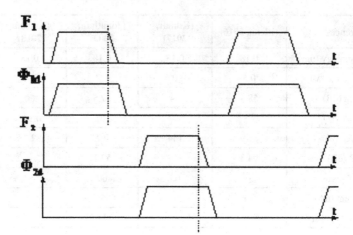

Switch Design

In general, the switch is one of the devices which generates a lot of imperfections in the SC circuit, such as gain error, offset and nonlinearity. The last type of error is very critical especially for the switches at the circuit input. Contrary to the NMOS and the PMOS switches, the CMOS switch allows the largest excursion of input signal. For this reason, our choice is directed towards the CMOS switch. In addition, a ghost transistor where its drain and source are short circuited, is used to absorb the charges injected by the analog switch M_n, then, the main switch charge injection impact is minimized (Figure 23) (Bob, 2012).

If the clocks phases are complementary and the transistor operates in the ohmic mode, then the equivalent on-resistance of the switch is given by:

$$R_{on} = R_{NMOS} / R_{PMOS} \tag{24}$$

Figure 23. Switch basic circuit with its ghost

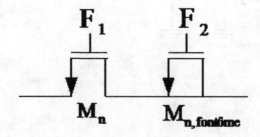

Where

$$R_{NMOS} = \frac{1}{\mu_n C_{ox} \left(\dfrac{W}{L}\right)_n \left(V_{dd} - V_{in} - V_{TN}\right)} \tag{25}$$

and

$$R_{PMOS} = \frac{1}{\mu_p C_{ox} \left(\dfrac{W}{L}\right)_p \left(V_{in} - V_{TP}\right)} \tag{26}$$

In order to make the switch on-resistance independent of input voltage, we choose:

$$\mu_n C_{ox} \left(\frac{W}{L}\right)_n = \mu_p C_{ox} \left(\frac{W}{L}\right)_p \tag{27}$$

More critical about the switch on-resistance is that it results in settling error. The first criterion to be satisfied after this preliminary choice is the C_s capacitor load rate. The time constant is equal to $\tau = R_{on} C_s$, it would be enough to choose the lowest possible resistance value to have the fastest settling of V_{in}. In other words, to have a fast settling, it would be enough that the W value is very high. Indeed, a second criterion must also be satisfied: the error generated by the charge injection and introduced to the V_{out} output voltage must be minimized and it is defined by:

$$\Delta V = \frac{WLC_{ox}\left(V_{dd} - V_{in} - V_{TN}\right)}{C_L} \tag{28}$$

This error is proportional to W. Therefore, a tradeoff between precision and settling speed is set. the CMOS switch use is able to minimize the charge injection when assuming that the charge quantity injected by the NMOS transistor is equal to that of the PMOS transistor. This solution seems judicious to reduce the charge injection and to improve the input dynamic. The adopted design method during this study consists of improving the settling precision by fixing the length of the two transistors at the minimal value of technology (0.18μm). Then the parameter W is computed to satisfy the settling time requirement.

Capacitors Sizing

The capacitors design has to take care of the TSMC technology design rules so that the capacitor mismatch is smaller than 0.5%. The values $C_s=0.01pF$ and $C_f=4pF$ meet this requirement.

Simulations Results

We simulate the SC integrator of Figure 21 for a 480 KHz switching frequency fs. Figure 24 shows a transient simulation where the integrator action for a constant input signal can be seen in the graph, which shown the differential output voltage. Figure 25 shows the results of a transient simulation using a 5KHz differential input signal.

CONCLUSION AND FUTURE EXTENSIONS

After presenting different OTAs structures, it has been noticed that the Telescopic OTA circuit is the most used circuit for low noise, low power consumption wireless communication system. This chapter presented a design of an ultra-low power telescopic OTA using the Bulk-Driven technique. This technique is adopted to overcome the threshold voltage limitation and to reduce the power consumption which is the main issue of the wearable wireless communication system. The proposed OTA architecture was optimized through the use of the PSO algorithm to

Figure 24. Differential output voltage

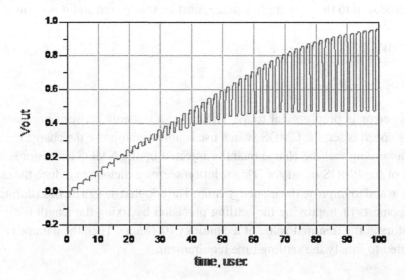

Figure 25. Differential output voltage at fs=480 KHz

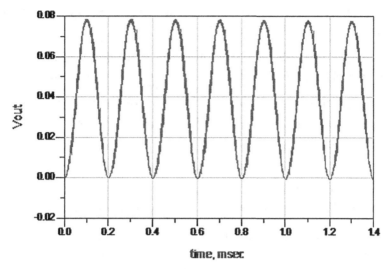

attain high performances in terms of DC gain and power consumption that satisfy the biomedical devices requirements. We have shown how the PSO program can be used to design and to optimize the Bulk-Driven telescopic OTA. This algorithm yielding globally optimal designs, is extremely efficient, and handles a wide variety of practical constraints. The method could be used, for example, to do full custom design for each op-amp in a complex mixed-signal integrated circuit; each amplifier is optimized for its load capacitance, required bandwidth, closed-loop gain, ..etc. To test the robustness of the optimized proposed OTA, it was used for the SC integrator implementation where the switches and the C_S and the C_f capacitors are carefully designed to attain a fast settling response. the authors conclude that the proposed OTA is also suitable for low power SC applications. The proposed telescopic bulk-driven OTA can be used for biomedical dispositive.

Future works would involve the investigation of the optimized proposed bulk-driven OTA in the Analog to Digital Converter (ADC) design which it is a critical block in the receiver channel for biomedical devices.

REFERENCES

Rajora, Zou, Guang Yang, Wen Fan, Yi Chen, Chieh Wu, ... Liang. (2016). A split-optimization approach for obtaining multiple solutions in single-objective process parameter optimizatio. *SpringerPlus*.

Abayomi, A. A., Ikuomola, A. J., Aliyu, O.A., & Alli, O. A., (2014). Development of a Mobile Remote Health Monitoring system – MRHMS. *IEEE African Journal of Computing & ICT*, *7*(4), 14-22,

Ahmadpour, A., & Torkzadeh, P. (2012). An Enhanced Bulk-Driven Folded-Cascode Amplifier in 0.18 μm CMOS Technology. *Circuits and Systems*, 187-191.

Ahmadreza, R. (1996). A 1-GHz CMOS RF front-end IC for a direct-conversion wireless receiver. *IEEE Journal of Solid-State Circuits*, *31*(7), 880–889. doi:10.1109/4.508199

Kilani, D., Alhawari, M., Mohammad, B., Saleh, H., & Ismail, M. (2016). An Efficient Switched-Capacitor DC-DC Buck Converter for Self-Powered Wearable Electronics. *IEEE Transactions on Circuits and Systems. I, Regular Papers*, 1–10.

Anisheh, S. M., & Shamsi, H. (2017). Two-stage class-AB OTA with enhanced DC gain and slew rate. *International Journal of Electronics Letters*, 4.

Cabrera-Bernal, E., Pennisi, S., Grasso, A. D., Torralba, A., & Carvajal, R. G. (2016). 0.7-V Three-Stage Class-AB CMOS Operational Transconductance Amplifier. *IEEE Transactions on Circuits and Systems*, *36*, 1807-1815.

Nye, W., Riley, D. C., & Sangiovanni-Vincentelli, A. L. (1988). DELIGHT.SPICE: An optimization-based system for the design of integrated circuits. *IEEE Transactions on Computer-Aided Design of Integrated Circuits and Systems*, *7*(4).

\Dobkin & Williams. (2012). *Analog Circuit Design Volume 2: Immersion in the Black Art of Analog Design*. Newnes.

Flandre, S. (1996). A gm/ID based methodology for the design of CMOS analog circuits and its application to the synthesis of a silicon-on-insulator micropower OTA. *IEEE Journal of Solid-State Circuits*, *31*(9), 1314–1319. doi:10.1109/4.535416

Hasançebi, O., & Kazemzadeh Azad, S. (2012). An Efficient Metaheuristic Algorithm for Engineering Optimization: SOPT. *International Journal of Optimization in Civil Engineering*, 479-487.

Hassan, K., & Shamsi, H. (2010). A sub-1V high-gain two-stage OTA using bulk-driven and positive feedback techniques. *European Conference on Circuits and Systems for Communications*.

Jonathan, R. J., Kozak, M., & Friedman, E. G., (2003). A 0.8 Volt High Performance OTA Using Bulk-Driven MOSFETs for Low Power Mixed-Signal SOCs. *IEEE International Systems-on-Chip, SOC Conference*.

Jung, W. (2005). *Op Amp Applications Handbook*. Newnes.

Karima, G. K., Hassen, N., Ettaghzouti, T., & Besbes, K. (2018). A low voltage and low power OTA using Bulk-Driven Technique and its application in Gm-c Filter. *Multi-Conference on Systems, Signals & Devices*, 429-434.

Kumar, M., Kiran, K., & Noorbasha, F. (2017). A 0.5V bulk-driven operational trans conductance amplifier for detecting QRS complex in ECG signal. *International Journal of Pure and Applied Mathematics*, *117*, 147–153.

Saidulu, B., Manoharan, A., & Sundaram, K. (2016). Low Noise Low Power CMOS Telescopic-OTA for Bio-Medical Applications. *Journal of Computers*, *5*(25).

Mainardi, Cerutti, & Bianchi. (2006). *The Biomedical Engineering Handbook, Digital Biomedical Signal Acquisition and Processing*. Taylor & Francis Group.

Matej, R. (2017). Design techniques for low-voltage analog integrated circuits. *Journal of Electrical Engineering*, *68*(4), 245–255. doi:10.1515/jee-2017-0036

Matej, R. M., Stopjakova, V., & Arbet, D. (2017). Analysis of bulk-driven technique for low-voltage IC design in 130 nm CMOS technology. International *Conference on Emerging eLearning Technologies and Applications (ICETA)*.

Michiel, S. M., van Roermund, A. H. M., & Casier, H. (2008). Analog Circuit Design: High-speed Clock and Data Recovery, High-performance Amplifiers, Power management. Springer.

Mohamed, K. (2013). Wireless Body Area Network: From Electronic Health Security Perspective. *International Journal of Reliable and Quality E-Healthcare*, *2*(4), 38–47. doi:10.4018/ijrqeh.2013100104

Roudjane, M., Khalil, M., Miled, A., & Messaddeq, Y. (2018). New Generation Wearable Antenna Based on Multimaterial Fiber for Wireless Communication and Real-Time Breath Detection. *Photonics*, *5*(33), 1–20.

Nikhil, Ranitesh, & Vikram. (2010). Bulk driven OTA in 0.18 micron with high linearity. *International Conference on Computer Science and Information Technology*.

Raj, N., Sharma, R. K., Jasuja, A., & Garg, R. (2010). A Low Power OTA for Biomedical Applications. *Journal of Selected Areas in Bioengineering*.

Raj, S. R., Bhaskar, D. R., Singh, A. K., & Singh, V. K. (2013). Current Feedback Operational Amplifiers and Their Applications. In Analog Circuits and Signal Processing. Springer Science & Business Media.

Ramirez-Angulo, J., Lopez-Martin, A.J., Carvajal, R.G., & Galan, J.A. (2002). A free but efficient class AB two-stage operational amplifier. *ISCAS*.

Ravelomanantsoa, A. (2015). *Deterministic approach of compressed acquisition and signals reconstruction from distributed intelligent sensors* (Thesis). University of Lorraine.

Samaher, J. (2017). Survey of main challenges (security and privacy) in wireless body area networks for healthcare applications. *Egyptian Informatics Journal, 18*(2), 113–122. doi:10.1016/j.eij.2016.11.001

Sameh, B., Sallem, A., Kotti, M., Gaddour, E., Fakhfakh, M., & Loulou, M. (2010). Application of the PSO Technique to the Optimization of CMOS Operational Transconductance Amplifiers. *IEEE International Conference on Design & Technology of Integrated Systems in Nanoscale Era*.

Sanchez-Sinencio, E., & Silva-Martínez, J. (2004). Continuous-Time Filters from 0.1Hz to 2.0 GHz. In *IEEE ISCAS*. Texas A & M University.

Galan, J. A., López-Martín, A. J., Carvajal, R. G., Ramírez-Angulo, J., & Rubia-Marcos, C. (2007). Super Class-AB OTAs With Adaptive Biasing and Dynamic Output Current Scaling. *IEEE Transactions on Circuits and Systems*.

Shant, K., & Mahajan, R. (2016). Techniques for the Improvement in the Transconductance of a Bulk Driven Amplifier. *Int. Journal of Engineering Research and Applications, 6*(2), 101–105.

Stopjakova, V., Rakus, M., Kovac, M., Arbet, D., Nagy, L., Sovcik, M., & Potocny, M. (2018). Ultra-Low Voltage Analog IC Design: Challenges, Methods and Examples. *Wuxiandian Gongcheng, 27*(1).

Vanichprapa, S., & Prapanavarat, C. (2010). Design of a simple OTA-based mixed-signal controller for adaptive control system of shunt active power filter. *WSEAS Transactions on Circuits and Systems, 9*(5), 305-314.

Wei, L. W. (2018). Improving Particle Swarm Optimization Based on Neighborhood and Historical Memory for Training. Information.

Yavari, M., & Shoaei, O. (2005). A Novel Fully-Differential Class AB Folded Cascode OTA for Switched-Capacitor Applications. *ICECS*.

Chapter 4
Enhancing System Reliability Through Targeting Fault Propagation Scope

Hakduran Koc
University of Houston, Clear Lake, USA

Oommen Mathews
University of Houston, Clear Lake, USA

ABSTRACT

The unprecedented scaling of embedded devices and its undesirable consequences leading to stochastic fault occurrences make reliability a critical design and optimization metric. In this chapter, in order to improve reliability of multi-core embedded systems, a task recomputation-based approach is presented. Given a task graph representation of the application, the proposed technique targets at the tasks whose failures cause more significant effect on overall system reliability. The results of the tasks with larger fault propagation scope are recomputed during the idle times of the available processors without incurring any performance or power overhead. The technique incorporates the fault propagation scope of each task and its degree of criticality into the scheduling algorithm and maximizes the usage of the processing elements. The experimental evaluation demonstrates the viability of the proposed approach and generates more efficient results under different latency constraints.

DOI: 10.4018/978-1-7998-1718-5.ch004

INTRODUCTION

Embedded technology solutions have become pervasive owing to the convenience they offer across a plethora of applications. As the computing environments running embedded applications are typically characterized by performance and power constraints, numerous energy-efficient performance-driven techniques have been proposed in the literature. In addition, since the accuracy of the results is the primary concern of design process, system reliability is also considered as another important optimization metric in any computing platform. Reliability of a system can be defined as the probability that the system generates its intended output in accordance with specifications for a given time period knowing that it was functioning correctly at the start time. Critical real time systems must function correctly within timing constraints even in the presence of faults. Therefore, the design of such systems in conjunction with fault tolerance is a challenge. The fault tolerance is defined as the ability of the system to comply with its specifications despite the presence of faults in any of its components (Ayav, Fradet, & Girault, 2006). With submicron process technologies, many faults have surfaced. Such faults can be categorized as a) permanent faults (e.g., damaged micro controllers or communication links), b) transient faults (e.g., electromagnetic interference), and c) intermittent faults (e.g., cold solder joint). Among them, transient faults are the most common one due to increasing complexity, transistor sizes, wafer misalignments, operational frequency, and voltage levels. The challenge is often to introduce fault-tolerance in an efficient way while maintaining performance constraints.

When reliability is improved using different techniques such as checkpointing or replication, there is invariably a performance overhead as well. In many cases, the penalty is additional off-chip memory access which may cause performance degradation and energy overhead. The situation is more serious in the case of chip multiprocessor-based environments wherein the multiple processors may try to simultaneously use the limited available off-chip bandwidth, resulting in bus contention problems thereby hampering the performance and causing more power consumption (Koc, Kandemir, Ercanli, & Ozturk, 2007; Koc, Kandemir, & Ercanli, 2010).

In this work, we propose a technique based on task recomputation in order to improve the system reliability without incurring any performance degradation. Given that an embedded application is represented using a task graph, the recomputation technique recomputes the result of a task using its predecessors whenever it is needed instead of making an explicit off-chip memory access if it is beneficial in terms of performance or energy. The proposed technique is an enhanced version of task recomputation handling different fault scenarios. The approach iteratively searches for idle time frames in available processing cores and assigns different tasks for recomputation based on criticality with respect to the task graph. The necessary condition for a task to be recomputed is that all its preceding tasks have been scheduled for execution and the outputs of such tasks are available in memory. Steps are then taken to make sure that the task is recomputed immediately prior to its requirement by successive tasks so memory is not occupied for a prolonged period. Each recomputation is made in accordance with the latency constraint and available resources so that the overall execution deadline is met.

The proposed approach involves a new fault-tolerant scheduling algorithm with two metrics namely, Fault Propagation Scope (FPS) and Degree of Criticality (DoC). Using these two variables, our technique strives to reduce the probability of fault propagation in the course of execution in a strategic manner. The roadmap of the approach encompasses the technique of recomputation to be applied to certain tasks based on their FPS and DoC values. FPS is an effective measurement in analyzing and evaluating the gravity of faults in the embedded systems. The approach has been used to perform experiments and tested with benchmarks to produce conclusive results which will be further explained as this chapter continues.

The hypothesis is that, instead of traditional fault tolerant approaches such as replication or checkpointing, a modified version of the recomputation can be used to improve the reliability in an efficient manner. The overall objective is to understand how to achieve robust scheduling by defining the scope for fault propagation in different task graphs. In this chapter, we present (1) two metrics, namely degree of criticality and fault propagation scope, required for designing our new scheduling algorithm, (2) the target architecture to exemplify the approach, (3) the reliability cost model that forms the mathematical foundation for reliability calculation, (4) an in-depth introduction to former methods of task recomputation, (5) the proposed algorithm to schedule the tasks in an efficient manner, and (6) an experimental evaluation of the proposed approach using different task graphs and benchmarks under different latency constraints.

PRELIMINARIES

Fault Propagation Scope

A data flow graph (DFG) is used to represent an embedded application in this chapter. A DFG is a directed acyclic graph consisting of vertices (representing operations) and edges (representing dependences between operations). An example DFG with six operations is shown in Figure 1. The edge from operation 0 to operation 1 indicates that operation 1 depends on operation 0; hence operation 1 cannot start executing until operation 0 completes its execution. Similarly, operation 3 can start executing after both operations 1 and 2 finish their executions.

Understanding the factors that determine the reliability of execution of operations/ tasks in a data flow graph is crucial in defining the methods to inculcate fault tolerance. Fault tolerance approaches can be broadly considered at different levels such as hardware, software, and high-level synthesis. Fault tolerance is introduced by employing system redundancy (addition of processing elements). By doing so, it is expected that in case of failures the tasks will be re-allocated (Lee, Kim, Park, Kim, Oh, & Ha, 2010) or be computed simultaneously on different processing elements (Briggs, Cooper, & Torczon, 1992). Such techniques may result in area and cost overheads and introduce an accelerated aging effect. When a system design is modified to include more processing elements, it may further result in the creation of hot spots (areas on the chip with higher on-chip temperature) (Xie, & Hung, 2006). Software techniques such as checkpointing and error correction codes may be useful for tolerating transient faults as well, but the associated overhead increases substantially as the number of transient faults rises. Moreover, in most cases such approaches are relevant only in the case of soft real-time applications and do not guarantee fault tolerance for a hard real-time system (Baunmann, 2005). Significant gains in reliability can be achieved when it is targeted at higher levels with an associated lower cost. Significant reliability improvements can only be achieved when reliability is treated as a first-class metric at the system level. Such an approach presenting an effective tradeoff between area, performance, and reliability is presented in (Tosun, S., Mansouri, N., Arvas E., Kandemir, M., Xie, Y., & Hung, W.L., 2005).

We define Fault Propagation Scope (FPS) in two ways: 1) for a task - the total number of successive tasks whose computation depends on a reliable execution of the task under consideration, and 2) for a task graph - the total number of edges in a task graph that are not fault-tolerant. A task with higher FPS value has priority over other tasks when implementing reliability measures in order to reduce the scope of the faults. For example, in Figure 1, if task 0 is not made fault tolerant, any error will force the re-execution of all other tasks (task 1, 2, 3, 4, and 5) because they are all dependent on the successful execution of task 0. So, FPS is 5 for task 0. FPS of the task graph is 7 because there are 7 edges in the graph which are not currently fault tolerant.

Degree of Criticality

Degree of Criticality (DoC) is defined as the longest distance of the task under consideration to the final task in a data flow graph. In addition to the FPS, DoC defines the importance of a specific task within the context of the entire task graph.

Figure 1. A sample DFG and Fault Propagation Scope

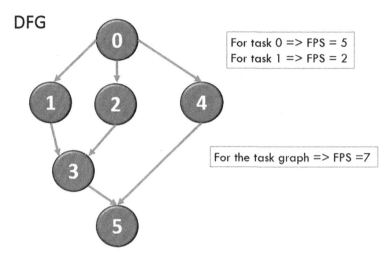

In Figure 2, DoC of 1 is 2 and DoC of task 0 is 3. Even though the tasks with higher DoC value are typically part of the critical path, our observations on various task graphs reveals that it is also possible for certain tasks to have higher DoC and not be a part of the critical path. It is important to consider such tasks in order to formulate a comprehensive fault tolerant mechanism. The proposed scheduling algorithm considers both DoC and FPS of each task in order to generate a reliable execution schedule.

Task Recomputation ·

The reliability improvement technique proposed in this chapter is based on task recomputation. The proposed technique recalculates the results of select tasks whenever their results are needed instead of making an off-chip memory access. For example, when task 3 is executing the results of task 1 and task 2 are needed. Instead of accessing memory for these results, task 1 and task 2 are recomputed and results are directly fed to task 3. This technique was first analyzed in the context of register allocation and further employed for different purposes such as improving performance, memory space consumption, or power/energy consumption. The overall execution latency is determined by the critical path, which is the path that takes the longest time to execute from source to sink node (Zhu, Melhem, & Mosse, 2004). Note that source node is task 0 and sink node is task 5 in Figure 2. Considering the fact that recomputations reduce the number of off-chip memory accesses (Koc, Kandemir, Ercanli, & Ozturk, 2007), the technique can be used to improve the performance and power consumption (Ziegler, Muhlfeld, Montrose, Curtis, O'Gorman, & Ross,

Figure 2. Degree of Criticality

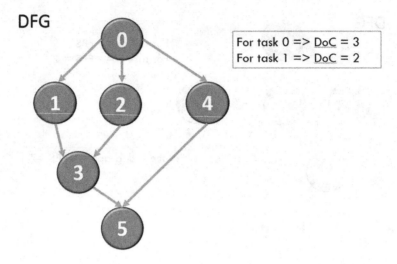

1996). In some cases, recomputations can be carried out during the idle times of the processing elements without incurring any performance overhead.

Now let us show how recomputation is performed using an illustrative example. We assume the technology library in Table 1 and the example task graph in Figure 3. The tasks are executed by a multi-core architecture with two processors, namely Processor 1 and Processor 2. The middle columns in table presents the reliability value of a task if it is executed by each available processor. The last column reports the execution latencies of the tasks. The latency constraint is given as 50 clock cycles. In this work, we utilize the algorithm developed by Tosun et al. (Tosun, S., Mansouri, N., Arvas E., Kandemir, M., Xie, Y. & Hung, W.L., 2005). The execution schedule is shown in Figure 3 on the right. As seen in the figure, task 1 is executed by Processor 2 and other tasks are executed by Processor 1.

Assuming a transient error affecting the execution of task 2, it is necessary to recompute the task and send the corrected result to task 4 in order to successfully execute the application. It is apparent from the schedule that Processor 2 is not fully utilized. In order to perform the recomputation of task 2, the idle time frame after the execution of task 1 on Processor 2 is used. The modified schedule with the recomputation of the task 2 is shown in Figure 4. Note that 2' represents the recomputation of task 2 by Processor 2. The increase in reliability is approximately 3%. It is quite possible to recompute task 0 as well on Processor 2 to further improve reliability.

RELATED WORK

Reliability improvement considering fault propagation scope and degree of criticality in the context of task recomputation was studied in (Mathews, Koc, & Akcaman, 2015). Various IC errors that adversely affect reliability can be broadly classified into two categories: extrinsic failures and intrinsic failures (Huang, Yuan, & Xu, 2009). Extrinsic failures are mainly caused by manufacturing defects whereas the intrinsic failures are caused by radiation effects. Intrinsic failures may be further subdivided into soft errors and hard errors. The initial approaches to enhance reliability were based on hardware replication. In (Chabridon & Gelenbe, 1995), the fault tolerance was based on the duplication of hardware resources and immediate comparison of the results. But this technique was not adequate to detect transients that are caused by the external event. A certain amount of temporal displacement

Table 1. Technology library for the tasks in Figure 3

Task	Reliability on Processor 1	Reliability on Processor 2	Execution Latency (Clock Cycles)
0	0.977	0.972	10
1	0.976	0.979	10
2	0.974	0.971	10
3	0.978	0.976	10
4	0.980	0.978	10

Figure 3. An example DFG and the corresponding execution schedule

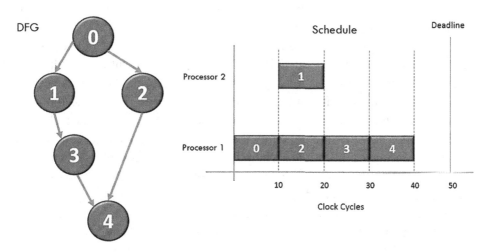

Figure 4. Modified schedule with the recomputation of task 2

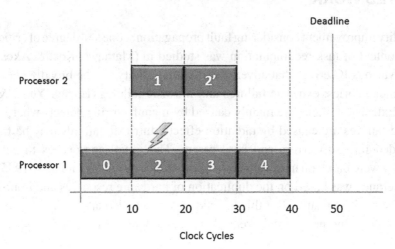

was introduced between redundant computations in order to realize a very high error detection coverage (Kopetz, Kantz, Grunsteidl, Puschner, & Reisinger, 1990). Even though such techniques were able to counter certain number of transient faults, it failed with the substantial increase in the number of the faults and hence became a costly approach. (Koc, Shaik, & Madupu, 2019) studied the modelling of a system consisting of a computation and control unit, a communication network, and physical elements such as sensors and actuators. They consider each component individually and use series and parallel configurations when calculating the reliability of the system. (Namazi, Safari, & Mohammadi, 2019) proposed a reliability-aware task scheduling algorithm for multicore systems with a quantitative reliability model. They try to keep the task replications and latency overhead at minimum while maintaining the requested reliability threshold and take communication overhead between processor cores into account.

In order to reduce the cost software replication (Chevochot, & Puaut, 1999), checkpointing with recovery (Xu, & Randell, 1996) and re-execution (Kandasamy, Hayes, & Murray, 2003) were considered. When these techniques were applied in a straightforward manner it would result in undesirable overheads and diminish performance. (Huang, Yuan, & Xu, 2009) introduced a fault-tolerant technique that dealt with a new lifetime-reliability aware task allocation and scheduling. An alternative to resource replication was introduced with the concept of task remapping wherein the tasks running on a failed processor are moved to another processor on detection of a failure with an additional cost of intensive compile-time computation and memory space [7]. Most of the related work in terms of task migration was primarily on reducing the migration cost (Gond, Melhem, & Gupta, 1996; Chean,

& Fortes, 1990; Manimaran, & Murthy, 1998; Suen, & Wong, 1992; Chang, & Oldham, 1995; Dave, & Jha, 1999). Nevertheless, it proved to be useful for multiple processor failures.

The energy-aware computation duplication approach was found to be effective in case of array-intensive applications and was efficient under the assumption that there are many idle processors (Chen, Kandemir, & Li, 2006). Multiple binding of the tasks while considering multiple objectives to design reliable datapaths was performed in order to improve the reliability (Glaβ, Lukasiewycz, Streichert, Haubelt, & Teich, 2007). The authors considered different objectives simultaneously. In (Agarwal, & Yadav, 2008) Agarwal and Yadav presented a system level energy aware fault tolerance approach for real time systems. (Akturk, & Karpuzcu, 2017) studied the tradeoff between communication and computation in terms of energy efficiency. The proposed technique eliminates an off-chip memory access if recomputing the value is energy-efficient.

The recomputation technique was studied first in the context of register allocation by Briggs et al. [14]. In their method, the execution is modeled in such a way that a register value is preferred to be rematerialized (recomputed) rather than storing and retrieving it from the memory. Kandemir et al. (Kandemir, Li, Chen, Chen, & Ozturk, 2005) employed recomputation to reduce the memory space consumption and in certain cases it also led to performance enhancement in applications involving memory-constrained embedded processing. Their approach basically recomputes the result of a code block when needed instead of storing it in memory. A detailed study on the relative advantages of the dynamic scheme of recomputation versus the static scheme was also covered in (Kandemir, Li, Chen, Chen, & Ozturk, 2005). (Safari, Ansari, Ershadi, & Hessabi, 2019) used task replication in order to satisfy reliability requirements and increase the Quality-of-Service in the context of mixed criticality systems. They target at improving the execution time overlap between the primary tasks and replicas. Memory space consumption of data-intensive applications was reduced using recomputation by Tosun et al. (Tosun, Kandemir, & Koc, 2006). Koc et al. (Koc, Kandemir, Ercanli, & Ozturk, 2007) further aimed at reducing the off-chip memory accesses in order to improve the performance of chip multi-processors. A recomputation based algorithm was proposed in (Tosun, S., Mansouri, N., Arvas E., Kandemir, M., Xie, Y. & Hung, W.L., 2005) to reduce the memory consumption in heterogeneous embedded systems. The outcomes of the algorithm were two extreme schedules with either the best performance or the best memory consumption. It was then modified to accept user input variables to come up with a final design with acceptable performance and memory requirements. In (Koc, Kandemir, & Ercanli, 2010) Koc et al. introduced a novel technique to exploit the large on-chip memory space through data recomputation. The technique was specifically applicable to the architectures that accommodate large on-chip software-managed memories. In such architectures, it was found efficient to recompute the value of an on-chip data, which is far from the processor, using the closer data elements instead of directly accessing the far data.

Some research has been devoted to comprehensive system optimization in the context of fault tolerance. For instance, Girualt et al. (Girualt, Kalla, Sighireanu, & Sorel, 2003) proposed a heuristic for combining several static schedules in order to mask fault patterns. Through active replication multiple failures were addressed in order to come up with an acceptable level of fault tolerance and performance constraints in (Kandemir, Li, Chen, Chen, & Ozturk, 2005). In the proposed approach, the technique of recomputation is utilized to improve reliability in a comprehensive manner without causing any performance overhead and simultaneously making sure that the processing elements are utilized efficiently. To our knowledge, there has not been any substantial work that targeted the critical nature of specific tasks and exploited such tasks to improve the reliability without creating performance overhead.

RELIABILITY COST MODELLING

The attainment of reliability is often cumbersome in complex systems. The system designers should consider different approaches and weigh their advantages and disadvantages before narrowing down to a preferred approach. Even then, all approaches should be based on certain foundational principles. In this section, such basic concepts surrounding the reliability calculation are discussed.

In performing reliability analysis of a complex system, it is almost impossible to treat the system in its entirety. The logical approach is to calculate the reliability of the entire system by factoring in the reliability of each component. Reliability of a given task can simply be defined as the probability that the task generates its intended output for a given time period knowing that it was functioning correctly at the start time. Reliability value varies from 0 to 1, where 1 represents 100% reliable output. Given that a task graph is used to represent an application, the tasks are performed in a series or parallel pattern depending on when their inputs are available during the course of execution. However, in order to ensure the successful execution of an embedded application, each and every operation has to execute successfully. Therefore, the original system is represented using a series configuration in terms of the reliability because the failure of any single component interrupts the path and causes the whole system to generate an erroneous output. The series reliability configuration is shown in Figure 5. The reliability of a single task is the probability of success of that task t, represented as P_i, each of these probability values is independent of the rest of the system. The reliability of the whole system, R, is the intersection of the probability of success of each task, which is expressed in Equation (1), and then, can be re-written as shown in Equation (2) (Shooman, 2002). Assuming that the failure rate is constant and is independent of their usage time, Equation (2) further simplifies to Equation (3), where R_s represents the system reliability, R_i is reliability of the task i and n is the number of tasks in the task graph.

$$R = P_s = P\left(x_1 x_2 x_3 \ldots x_n\right) \tag{1}$$

$$P_s = P\left(x_1\right) P\left(x_2\right) P\left(x_3\right) \ldots P\left(x_n\right) \tag{2}$$

$$R_s = \prod_{i=1}^{n} R_i \tag{3}$$

The series configuration forms the lowest bound in terms of reliability. The straightforward approach to improve reliability is by adding redundancy (i.e., add redundant tasks that would take over and perform computation in case of failure of the primary task). In this work, the redundancy is added by using the task recomputation in the idle time frames of the processors. It results in creation of multiple execution paths, which increases the probability of successful execution. The reliability of a purely parallel system, such as the one represented in Figure 6, can be expressed as in Equation (4), which further reduces to Equation (5).

$$R = P_s = P\left(x_1 + x_2 + x_3 + \ldots + x_n\right) \tag{4}$$

$$R_s = 1 - \prod_{i=1}^{n}\left(1 - R_i\right) \tag{5}$$

Figure 5. Series reliability configuration

Figure 6. Parallel reliability configuration

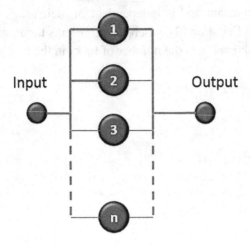

By transforming the series configuration to a parallel configuration using task recomputation, it is possible to obtain alternate execution paths. This allows a successful execution in case the initial computation path fails. The reliability of the system after applying the proposed recomputation (hybrid recomputation) technique is determined using a combination of series and parallel configurations. In other words, an original task and its recomputed version is a parallel subsystem, which is connected to the rest of the system. The reliability of the recomputed task is in series with the recomputed predecessors and together is in parallel with the original task.

The hybrid recomputation considers two different modes of possible recomputations, namely reprocessing and replication. The mode of execution in reprocessing follows a similar approach in (Eles, Izosimov, Pop, & Peng, 2008). Once a fault is detected, a fault tolerance mechanism is invoked to handle this fault. The simplest form to recover is re-execution (reprocessing). In such a scenario, the process is re-executed again on the same processing element. Since recomputation does not involve any time to restore the order of operations, there is no case for recovery overhead and thus the performance remains unaffected. The mode of reprocessing is illustrated Figure 7. A failure during the execution of the task 1 is overcome by subsequent recomputation in the same processor (Processor 1). Even though this mode of recomputation helps in cases where the radiation might cause transient faults in the memory, there are certain obvious limitations. Firstly, it results in poor utilization of the processing elements. Secondly, since spatial redundancy is not obtained in this case, it does not result in highly improved reliability. Finally, continuous usage of a specific processor may lead to the creation of hot spots and accelerated ageing effects.

In the replication mode, the recomputation is performed using another processing element as shown in Figure 8. (i.e., task 1, which is originally scheduled on Processor 1, is recomputed using Processor 2). Replication can be further subdivided into two categories: active and passive (Pop, Izosimov, Eles, & Peng, 2009). Active replication is where the recomputation happens simultaneously on another processing element. In case of passive replication, the alternate processing elements are used only in case of failures. Active and passive replication leads to spatial redundancy; hence, it produces better reliability results.

Hybrid recomputation combines both modes to develop an effective approach and it is applied based on the specific nature of a task graph. An analysis of different task graphs shows that the performance varies based on the modes of recomputation. The conditions that determine the selection of a specific mode are basically the availability of an idle processing element and the performance constraints. The hybrid mode of recomputation results in highly efficient utilization of the processing elements thereby reducing the possibility of idle time frames.

Figure 7. Recomputation in the reprocessing mode

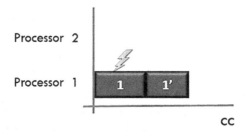

Figure 8. Recomputation in the replication mode

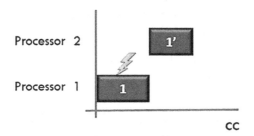

ENHANCING RELIABILITY BY TARGETING FAULT PROPAGATION SCOPE

The proposed algorithm integrates the concepts of FPS and DoC and takes different modes of recomputation into account to achieve the best possible results in terms of reliability and performance.

Target Architecture

The target architecture utilized in this work for different experiments is a multi-core embedded system as shown in Figure 9. It comprises of many processors that have different reliability, latency and area values and a synchronization/communication logic. Each task in a task graph can have different implementation options available on each processor with different reliability and latency metrics. In addition to the private L1 memory of each processor, the system has a shared on-chip and an off-chip memory. The processors' access time increases as it goes to the off-chip DRAM. Please note that an off-chip memory access is avoided when performing recomputations. Given this architecture, along with a technology library, a task graph of an application and performance deadline, the proposed approach produces a schedule with allowable recomputations using a certain number of processors.

Illustrative Example

In this work, both modes of recomputation are taken into account. Faults can occur both with the processor logic as well as the memory. FPS and DoC are considered while modifying the schedule. This work targets data-intensive applications that usually consists of loops and multi-dimensional arrays. In such applications, the total dataset size (including scalars, arrays and program code) are heavily dominated by array variables. Consider the code fragment shown below assuming array $T[$ $]$ is initialized earlier in the code. For the sake of simplicity, let us assume that the loop has one nest, $J = 100$, and the arrays $S[$ $]$, $W[$ $]$, $U[$ $]$, $F[$ $]$, $K[$ $]$, $A[$ $]$, $Q[$ $]$ & $K[$ $]$ are one-dimensional.

```
        begin
        for i:= 1 to J do
0           W[i] = T[i + 4] + 2;
1           S[i] = W[i] + 6;
2           U[i] = W[i] * 2;
3           F[i] = W[i] + 9;
```

Figure 9. Target architecture

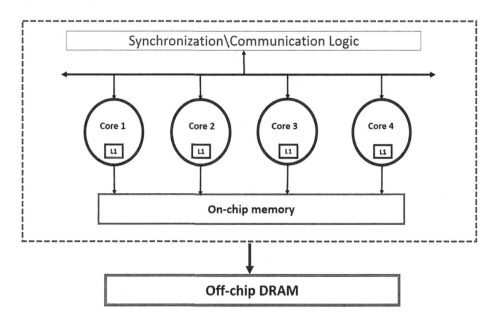

$$4 \qquad K[i] = \frac{U[i]}{4} + S[i]$$

$$5 \qquad A[i] = \frac{S[i]}{4};$$

$$6 \qquad Q[i] = K[i] * \left(F[i] + A[i]\right);$$

```
      end
```

The corresponding simplified version of the task graph is shown in Figure 10. The technology library for the task in Figure 10 is shown in Table 2. The scheduling technique developed in (Tosun, S., Mansouri, N., Arvas E., Kandemir, M., Xie, Y. & Hung, W.L., 2005) is used for different comparisons. The algorithm tries to allocate the tasks to the most reliable processors as soon as possible so that maximum slack is obtained. Owing to the varying execution latencies of the tasks, processors invariably become under-utilized. The corresponding initial schedule is shown in the Figure 11.

The schedule in Figure 11 shows that Processor 2 is idle most of the time. Note that FPS of the task graph is 9, which is the worst case without any redundancy. FPS of task 0 is 6 (i.e., if a fault occurs during the execution of task 0, it will get propagated to all other tasks in the graph). The reliability calculated using Equation (3) is 0.8662. The execution schedule after applying hybrid recomputation is shown in Figure 12. As seen in the figure, the tasks 0, 1, 2, and 3 are recomputed and they

Figure 10. Task graph representation of the example code fragment

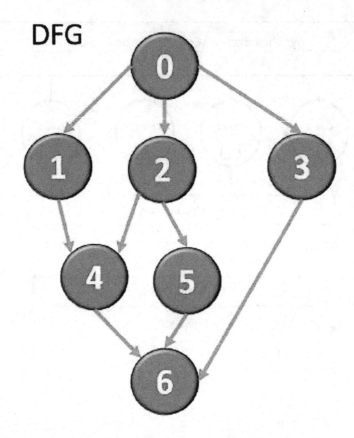

Table 2. Technology library for the tasks in Figure 10

Task	Reliability on Processor 1	Reliability on Processor 2	Execution Latency (Clock Cycles)
0	0.979	0.973	10
1	0.966	0.975	8
2	0.980	0.978	8
3	0.989	0.985	14
4	0.978	0.983	10
5	0.974	0.971	6
6	0.978	0.973	10

Figure 11. The initial execution schedule for the task graph in Figure 10

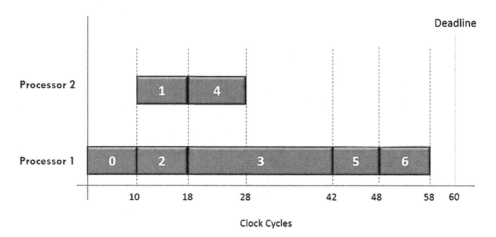

are shown as 0', 1', 2', and 3', respectively. Processor 2 becomes properly utilized in most of the time frames during which it was previously idle. Recomputation happens just before the succeeding task requires a result. For instance, recomputations of task 1 and task 2 are performed just before task 4 starts executing. An advantage of such a delayed form of recomputation is to use the on-chip memory space (or registers) only for a minimum time period. Recall that recomputation does not involve an off-chip memory access and hence the required inputs must be present in the nearby on-chip memory.

The reliability thus obtained after the hybrid recomputation is 0.9347 with an improvement of 7.9%. In addition to that, the FPS of the task graph is now reduced to from 9 to 2, thereby ensuring that the scope for the faults to propagate is considerably reduced. Another important observation is that the performance is unaffected (time taken for all the executions is less than the deadline), which is highly advantageous in the case of critical embedded systems.

Details of the Proposed Algorithm

The approach is designed for embedded systems running data-intensive applications. Considering multiple optimization metrics concurrently, the approach strives to improve the reliability of a system in a balanced and efficient manner. Data recomputation is chosen on the premise that an off-chip memory access takes many clock cycles; and hence it is better to access the required operands from on-chip memory and recompute the requested data. However, in order to perform the recomputation, certain criteria need to be followed. To begin with, for each of the

Figure 12. The execution schedule with recomputations

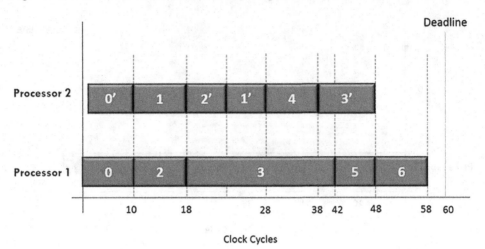

tasks in a task graph, the corresponding FPS and DoC values are determined. For instance, consider the task graph in Figure 10. The FPS values of tasks 0 to 6 are {6,2,3,1,1,1,0} and the DoC values are {3,2,2,1,1,1,0}, respectively. The precedence for recomputation is based on the decreasing order of these values. Hence, as seen in Figure 11, the first task on which the recomputation is performed is 0. One caveat to this approach would be in the selection of a task. When there are multiple tasks with the same DoC, but different FPS values, it might become an issue. For example, consider tasks 1 and 2. $DoCT1 = 2$ and $DoCT2 = 2$ but $FPST1 = 2$ and $_{FPS}T2 = 3$. In this case, task 2 which has a higher FPS is chosen for recomputation before task 1 as shown in Figure 11. When considering both FPS and DoC, FPS value of a task is given more precedence because the tasks with higher value of FPS are more critical to the successful execution of the entire application. After constructing the task graph from the given inputs, it sorts the tasks based on their FPS andDoC values. The individual tasks are scheduled in a reliability-centric approach in order to obtain the maximum slack so that maximum possible number of recomputations are performed. In doing so, the processor cores are efficiently utilized to its fullest capacity while ensuring that the performance remains unaffected.

Now let us discuss the proposed algorithm to perform the hybrid recomputation in detail. The pseudocode of the proposed heuristic algorithm is given below. The algorithm takes the raw task graph in an XML format, along with the technology library (also in XML format), and the performance bounds (*Lmax*) as inputs. The outputs include the modified task graph with recomputations, reduction of FPS (*FPSr*), the new latency bound (*Lnew*) and the reliability improvement gain (*RELi*). After reading the input task graph Gs (T, E) from lines [4-8], $DoCT$ and $FPSTi$ of each task from $i = 1$ to_n is calcula$_{ted.}$ In the same loop, the tasks are assigned to the processor core that has the highest reliability for the particular task. In line 7, the recomputation mode of the task is decided. For example, in Figure 11, the mode for task 0 is reprocessing. In line 9, the subroutine sorting the tasks based on *FPS* and *DoC* values is called. Again, in the previous example, after sorting, task 2 comes before task 1 in the queue for recomputation because *FPST2*> *FPST1* even though *DoCT1= DoCT2*. In the next few lines $_{[1}$0-12], e$_{ach}$ task is checked fo$_{r t}$he statu$_{s o}$f schedule of its preceding tasks. If all the preceding tasks are already scheduled, then the tasks that are not set for recomputation is determined. Note that the tasks that are eligible for recomputation *TRcompe* are sorted beforehand using the subroutine, *Sort_Tasks(DoCT, FPS$_{T)}$*. $_{In lines}$ [14-17], a feasibility check is performed on the different possible ti$_{me}$-frames $_{fo}$r recomputation of each eligible task *TRcompe* that are chosen for recomputation from 1 to r. The time frame is determined using the co$_{ndition in E}$quation (6):

$$\Omega Rcompe j + Lnew > Lmax \text{ (6)}$$

where $\Omega Rcompe j$ is the cost of the recomputation in terms of the number of $_{clock cycle}$s neede$_{d to a}$ccess $_{the o}$n-chip memor$_{y. The condi}$tion checks whether the recomputation of the task *Tj* in a specific timeframe exceeds the given maximum latency bound, *Lmax*. If the condition is satisfied, the particular tas$_k$ is set for recomputation and the schedule is modified accordingly [l$_{ine 21}$]. In line 26, the total Reliability, *Rtotal* and FPS of the graph, *FPSG* is calculated. Then, the given task graph is scheduled and the reliability improvement (*RELi*) a$_{nd the red}$uction in FPS (*FPSr*) are both $_d$etermined in percent.

```
ALGORITHM: Scheduling with Hybrid Recomputation
1: Input          : Gs(T, E), technology library, and
performanc_e (Lmax) bound
 2: Output         : Modified schedule with recomputations,
FPSr, RELi
 3: Read DFG and con_struct T = {T1, T2, …, Tn}
 4: for i = 1 to n do
 5:        Calculate DoCTi and FPSTi
 6:          Assign each task to processing ele_me_nt based _on
reliability
 7:         Assign MRcomp _to ob_tain Sma_x
 8: end for
 9: Sort_Tasks(DoCT, FPST)
10: for i = 1 _to n_ do
11:    _if T_pred = {T1, T2, …, TP} are scheduled then
12:     Determine tasks with status IsRecomp_=false and_ construct
{T_Rcompe}
```

```
13:    end if
14:    for (j = 1 to r) try recomputation
15:       During execution of actual t_ask, Ta
16:       After execution of actual task, Ta
17:       After execution of all recomputations assigned to the
processor
18:          if QRcompe j + Lnew > Lmax then
19:             try next TRcompej
20:          else
21:             Set IsRecompTRcompej = true and modify Gs(T, E)
22:          end if
23:       end for
24:    Schedule task Ti and update Lnew
25: end for
26: Calculate Rtotal and FPSG
27: Schedule Gs (T, E)
28: Calculate the FPSr and RELi
29: Generate output DFG showing all recomputations and
    start time and end time of all tasks
```

EXPERIMENTAL RESULTS

In order to evaluate the efficacy of the proposed approach, an experimental evaluation using different benchmarks are performed. The experiments are conducted using benchmarks programs and the task graphs automatically generated by the tool TGFF. The number of edges in the graphs ranges from 10 to 80. The algorithm is implemented using C# and executed on an Intel Core(TM) i5-4200 CPU with a RAM of 4.00 GB. The number of tasks in each graph is randomly varied and the approach is evaluated against all possible scenarios. The target architecture is similar to the one described earlier.

The results of the first set of experiments are shown in Table 3. The first column gives the label of the tasks, column two reports the number of edges in each task graph that was generated by TGFF. Column 3 presents the reliability value that is computed using the reliability-centric method described in (Tosun, S., Mansouri, N., Arvas E., Kandemir, M., Xie, Y. & Hung, W.L., 2005). In columns 4 to 8, the reliability values obtained using the proposed recomputation approach under different latency constraints is given (i.e., no latency overhead in column 4, 5% latency overhead in column 5, and so on). In column 8, the constraints placed on latency are taken off in order to know the upper limit of the reliability that can possibly be achieved using the proposed approach. The next five columns show the corresponding improvement in reliability. As seen in the table, using the same latency bounds, the proposed approach improves the reliability in the range of 5% to 16%, which is considerable improvement over the previous approaches without even affecting the performance.

In addition, the following observations can be made on this first set of results. Firstly, the proposed algorithm provides an effective utilization of available processing elements. When a small performance degradation is allowed, the approach generates significant reliability improvements. For example, in case of the task graph s05 a latency increase of just 5% result in the increase of reliability improvement by over 6%. Secondly, there are cases where an increase in the latency does not necessarily have no or marginal impact on the reliability. It can be seen in case of the task graph, s00 where the latency increase of 5% and 10% result in the same reliability value. This occurs because even if the latency is increased, there is not enough time frame available to perform more recomputations. Thirdly, the improvements obtained in the reliability can be bettered even further using the triplication (performing recomputation twice) of the highly critical tasks. However, that was not attempted because the main objective was to reduce the FPS along with increasing the reliability. In order to achieve that objective, the approach would involve the recomputation of as many tasks (with high DoC and large FPS) as possible within the available time frame. In certain scenarios, it was observed that even with an increase in the given latency, the reliability decreases marginally. It is because even when the focus is on those tasks with high FPS and DoC, there are other tasks which are assigned to the processors with low reliability. Not considering such tasks for recomputation adversely affects the overall reliability. However, such reductions in reliability is found to be negligible and thus not taken into consideration. Moreover, the cost of implementing fault-tolerant mechanisms for less critical tasks is very low hence it is better to always focus on the ones that are more critical as far as the overall schedule of the task graph is concerned.

Table 4 shows the percentage reduction in FPS using the approach for the same set of task graphs as in Table 3. The effective FPS of an entire task graph is analyzed in terms of the edges that is not fault tolerant. Please note that the number of edges in a task graph is will higher than the number of tasks. Hence, even if a small number of tasks are set for recomputation, FPS can be substantially reduced. For instance, in case of the task s06, a latency increase of just 5% reduces its FPS by 26% (with the same latency it was about 3.8%) because now 12 more edges are made fault tolerant. Based on the results obtained, it could be stated that FPS owing to its thus proven importance becomes a part of the metrics aimed at reducing the faults in task graphs. It is further based on the fact that the addition of FPS to the reliability approaches results in comprehensible and more efficient fault-tolerance improvements.

Table 3. Reliability improvements using the proposed approach for the graphs automatically generated by TGFF

Tasks	Edges	Tosun et al. 2007	Reliability						% Improvement in Reliability						Latency with Normal Scheduling	Fault Propagationless Latency
			Same Latency	Latency 5%	Latency 10%	Latency 25%	No Latency Constraint	Same Latency	Latency 5%	Latency 10%	Latency 25%	No Latency Constraint			
s00	7	0.923	0.957	0.958	0.958	0.9849	0.9849	3.673	3.770	3.770	6.706	6.706	42	53
s01	13	0.873	0.889	0.889	0.903	0.9221	0.942	1.810	1.810	3.436	5.624	7.847	61	84
s02	15	0.753	0.776	0.807	0.825	0.8578	0.987	3.054	7.118	9.588	13.918	31.089	72	124
s03	17	0.836	0.872	0.894	0.896	0.931	0.947	4.306	6.938	7.213	11.364	13.218	92	124
s04	21	0.735	0.774	0.792	0.813	0.8680	0.985	5.361	7.687	10.612	18.095	34.068	98	166
s05	44	0.561	0.615	0.648	0.683	0.7280	0.826	9.626	15.508	21.783	29.768	47.237	207	287
s06	52	0.533	0.582	0.589	0.627	0.6990	0.846	9.193	10.507	17.636	31.144	58.724	195	300
s07	68	0.431	0.503	0.541	0.591	0.675	0.798	16.787	25.702	37.288	56.652	85.280	337	478
s08	81	0.339	0.396	0.429	0.444	0.554	0.754	16.987	26.609	31.049	63.604	122.747	416	636

Table 4. Reduction of FPS using Hybrid Recomputation

Tasks	Edges	FPS					% Reduction in FPS				
		Same Latency	Latency 5%	Latency 10%	Latency 25%	No Latency Constraint	Same Latency	Latency 5%	Latency 10%	Latency 25%	No Latency Constraint
s00	7	3	2	2	0	0	57.143	71.429	71.429	100.000	100.000
s01	13	9	9	7	3	0	30.769	30.769	46.154	76.923	100.000
s02	15	12	11	10	8	0	20.000	26.667	33.333	46.667	100.000
s03	17	12	8	8	3	0	29.412	52.941	52.941	82.353	100.000
s04	21	16	16	14	10	0	23.810	23.810	33.333	52.381	100.000
s05	44	36	30	23	12	0	18.182	31.818	47.727	72.727	100.000
s06	52	50	38	38	20	0	3.846	26.923	26.923	61.538	100.000
s07	68	67	53	51	15	0	1.471	22.059	25.000	77.941	100.000
s08	88	84	74	58	30	0	4.545	15.909	34.091	65.909	100.000

The next set of experiments involves the data-intensive benchmarks, which are different types of filters: FIR, EW, DE, and ER. Table 5 presents the results of the experimental evaluation. The column labels follows the same order in Table 3. As seen, the proposed approach improves the reliability in the range of 38% without causing any performance overhead. Through these experiments it is identified that when the cost of the recomputation is reduced, better results are obtained using hybrid recomputation in comparison to the previous approaches.

The final set of experiments investigates the effect of the proposed approach when the volume of the computation is increased. The results are summarized in

Table 5. Hybrid recomputation using the benchmarks

Benchmark Codes	Tosun et al. 2007	Reliability					% Improvement in Reliability					Latency with Normal Scheduling	Fault Propagationless Latency
		Same Latency	Latency 5%	Latency 10%	Latency 25%	No Latency Constraint	Same Latency	Latency 5%	Latency 10%	Latency 25%	No Latency Constraint		
FIR Filter	0.691	0.714	0.730	0.735	0.784	0.973	3.284	5.627	6.321	13.410	40.749	123	204
EW Filter	0.662	0.711	0.723	0.750	0.820	0.914	7.402	9.215	13.293	23.867	38.066	152	216
DE Solver	0.863	0.896	0.903	0.907	0.930	0.949	3.800	4.611	5.074	7.739	9.940	71	99
AR Filter	0.632	0.680	0.708	0.719	0.772	0.967	7.595	12.025	13.734	22.152	53.006	128	217

Figure 13. Reliability improvements in large task graphs

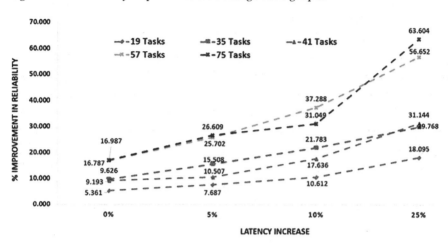

Table 6. Specifications for the Scheduler

Specification	Description
Task Graph	Any type of task graphs regardless of # of tasks or edges
# Tasks	No Restriction
# Edges	No Restriction
Algorithms	• Base Scheduling Approach • Recomputation in Replication Mode • Recomputation in Reprocessing Mode • Hybrid Recomputation
Input and Output File Format	XML
# Processors	2 (can be extended)
Output Parameters	• Reliability • FPS • Latency • % Improvement in Reliability • % Reduction in FPS • Start Time and End Time of all the tasks • Tasks set for recomputation
Lines of Code	<5000
Windows Application	WPF 4.5
Programming Language	C# 4.5
IDE	Microsoft Visual Studio 2013

Figure 13, where five large task graphs are subjected to hybrid recomputation. In case of large and intensive computations, this approach yields better results. This is owing to the fact that, with the increase in the number of tasks, usual scheduling approaches result in a poor utilization of the processing elements. However, our approach efficiently utilizes such idle time frames to perform recomputations. As the number of tasks were increased to 75, a substantial improvement in the reliability 16.98% was observed even with the same latency constraints.

Now let us describe the electronic design automation tool, named Scheduler, developed for scheduling different task graphs. The scheduler was implemented using Windows Presentation Foundation (WPF) and programmed using C#. WPF provides developers with a unified programming model for building rich Windows smart client user experiences that incorporate UI, media, and documents (Anonymous, 2014). The IDE used for developing the tool was Microsoft Visual Studio 2013. The specifications of the tool is presented in Table 6.

In this section, a detailed overview of novel fault-tolerant scheduling mechanism using hybrid recompuation was discussed. By targeting the fault propagation scope and degree of criticality, hybrid recomputation aims to enhance the reliability in an efficient manner. The added advantage of the technique is the fact the performance remains unaffected. Another important outcome is the excellent fault coverage, which is ensured by the effective utilization of the available processing elements.

CONCLUSION

In this chapter, a novel technique called hybrid recomputation was introduced to improve the reliability of embedded applications even without causing any performance overhead. By incorporating essential concepts such as Degree of Criticality and Fault Propagation Scope, it became possible to develop an effective method to enhance the reliability and simultaneously reduce the scope for further fault propagation. The underlying concept is to target the tasks that have higher impact on the overall successful execution of the application. Targeting such tasks, which are critical for reliable execution of an application, results in better fault coverage. The approach is validated using the benchmarks and the task graphs generated by TGFF. With the increase in the volume of computations, the approach is able to generate even better results both in terms of improved reliability and reduced FPS.

REFERENCES

Agarwal, S., & Yadav, R. S. (2008). System level energy aware fault tolerance approach for real time systems. IEEE Region 10 Conference, TENCON, 1-6.

Akturk, I., & Karpuzcu, U. R. (2017). AMNESIAC: Amnesic Automatic Computer. Proceedings of the Twenty-Second International Conference on Architectural Support for Programming Languages and Operating Systems, 811-824.

Anonymous. (2014). MSDN Windows Present*ation Foundation. Retrieved from* http://msdn.microsoft.com/en-us/library/vstudio/ms754130.aspx

Ayav, T., Fradet, *P., & Girault, A. (2006). Implementing Fault Tolerance in Real-Time Systems by Automatic Program Transformations. Proceedings of t*he 6th ACM & IEEE International *conference on Embedded software,* 205-214. 10.1145/1176887.1176917

Baunmann, R. (2005). Soft errors in advanced computer systems. IEEE Transactions on Design & Test of Computers, 22(3), 258-266.

Briggs, P., Cooper, K. D., & Torczon, L. (1992*). Rematerialization. Proceedings of the 92nd Programming Language Design and I*mplementation.

Chabridon, S., & Gelenbe, E. (1995). Failure detection algorithms for a reliabl*e execution of parallel programs. Proc. Internationa*l Symposium on Reliable Distributed Systems, 229-238. 10.1109/RELDIS.1995.526230

Chang, H. W. D., & Oldham, W. J. B. (1995). Dynamic task allocation models for large distributed computing systems. IEEE Transactions on Parallel and Distributed Systems, 6(12), 1301–*1315. doi:10.1109/71.476170*

Chean, M., & Fortes, J. (1990). The full-use-of-suitable-spares (FUSS) approach to hardware reconfiguration for fault-tolerant processor arrays. IEEE Transactions on Computers, 39(4), *564–571. doi:10.1109/12.54851*

Chen, G., Kandemir, M., & Li, F. (2006). Energy-Aware Computation Duplication for Improving Reliability in Embedded Chip Mulitprocessors. Asia and South Pacific Conference on Design Automation, 134-139.

Chevochot, P., & Puaut, I. (1999). Scheduling Fault-Tolerant Distributed Hard-Real Time Tasks Independently on the Replication Strategies. 6th Int. Conf. on Real-Time Comp. Syst. And Applications, 356-363. *10.1109/RTCSA.1999.811280*

Dave, B. P., & Jha, N. K. (1999). COFTA: Hardware-Software Co-Synthesis of Heterogeneous Distributed Embedded System for Low Overhead Fault Tolerance. IEEE Transactions on Comp*uters, 48(4), 417–441. doi:10.1109/12.762534*

Eles, P., Izosimov, V., Pop, P., & Peng, Z. (2008). Synthesis of Fault-Tolerant Embedded Systems. Design, Automation and Test in Europe, 1117-1122.

Girualt, A., Kalla, H., Sighireanu, M., & Sor*el, Y. (2003). An Algorithm fo*r *Au*tomatically Obtaining Distributed and Fault-Tolerant Static Schedules. Int. Conf. on Dependable Syst. And Netw., 159-168. 10.1109/DSN.*2003.1209927*

Glaß, M. Lukasiewycz, M., Streichert, T., Haubelt, C., & Teich, J., (2007). Reliability Aware System Synthesis. Design, Automation & Test in Europe Conference & Exhibition, 1-6.

Gon*d, C., Melhem, R., & Gupta, R. (1996).* Loop transformations for fault detection in regular loops on massively parallel systems. IEEE Transactions on Parallel and Distributed Systems, 7(12)*, 1238–1249. doi:10.1109/71.553273*

Huang, L., Yuan, F., & Xu, Q., (2009). Lifetime Reliability-Aware Task Allocation and Scheduling for MPSoC Platform. Design, Automation and Test in Europe Conf*erence & Exhibition, 51-56.*

Kandasamy, N., Hayes, J. P., & Murray, B. T. (2003). Dependable Communication Synthesis for Distributed Embedded Systems. IEEE Transactions on Computers, 52(5), 113–125. doi:10.1109*/TC.2003.1176980*

Kandemir, M., Li, F., Chen, G., Chen, G., & Ozturk, O. (2005). Studying storage recomputation tradeoffs in memory-constrained embedded processing. Proc. of the Conf. on De*sign, Automation and Test in Eu*rope, 1026-1031 10.1109/DATE.2005.285

Koc, H., Kandemir, M., & Ercanli, E. (2010). Exploiting large on-chip memory space through data recomputation. IEEE International SOC Conference, 513-51*8. 10.1109/SOCC.2010.5784683*

*Koc, H., Kandemir, M., Erca*nli, E., & Ozturk, O. (2007). Reducing Off-Chip Memory Access Costs using Data Recomputation in Embedded Chip Multi-processors. Proceedings of De*sign and Automation Conference, 224-229.*

Koc, H., Shaik, S. S., & Madupu, P. P. (2019). Reliability Modeling and Analysis for Cyber Physical Systems. Proceedings of IEEE 9th Annual Computing and Communication Workshop and *Conference, 448-451. 10.1109/CCWC.2019.8666606*

Koc, H., Tosun, S., Ozturk, O., & Kandemir, M. (2006). Reducing memory requirements through task recomputation in emb*edded systems. Proceedings of IEEE ComputerSociety Annual Symposium on VLSI: Emergi*ng VLSI Technologies and Architecture. 10.1109/ISVLSI.2006.77

Kopetz, H., Kantz, H., Grunsteidl, G., Puschner, P., & Reisinger, J. (1990). Tolerating Transient Faults *in MARS. 20th Int. Symp. On Fault-Tolerant Computing, 466-473.*

Lee, C., Kim, H., Park, H., Kim, S., Oh, H., & Ha, S. (2010). A Task Remapping Technique for Reliable Multi-Core Embedded Systems. Int. Conference on CODES/ ISSS, 307-316. 10.1145/1878961.1879014

*Manimaran, G., & Murt*hy, C. S. R. (1998). A fault-tolerant dynamic scheduling algorithm for multiprocessor real-time systems and its analysis. IEEE Transactions on *Parallel and Distributed Sys*tems, 9(11), 1137–1152. doi:10.1109/71.735960

Mathews, O., Koc, H., & Akcaman, M. (2015). Improving Reliability through Fault Propagation Scope in Embedded Systems. Proceeding*s of the Fifth International Conference on Digital In*formation Processing and Communications (ICDIPC2015), 300-305. 10.1109/ICDIPC.2015.7323045

Namazi, A., Safari, S., & Mohammadi, S. (2019). CMV: Clustered Majo*rity Voting Reliability-Aware Task Scheduling for Multicore Real-Time Systems. IEEE Transactions on Reliability, 68*(1), 187–200. doi:10.1109/TR.2018.2869786

Pop, P., Izosimov, V., Eles, P., & Peng, Z. (2009). Design Optimization of Time- and Cost Constrained Fault-Tolerant Embedded Systems With C*heckpointing and Replication. IEE*E *Tr*ans. on VLSI Systems, 389-401.

Pop, P., Poulsen, K. H., Izosimov, V., & Eles, P. (2007). Scheduling and Voltage Scaling for Energy/Reliability Trade-offs in Fault-Tolerant Time-Triggered Embedded Systems. 5th *International Conference o*n CODES+ISSS, 233-238. 10.1145/1289816.1289873

Safari, S., Ansari, M., Ershadi, G., & Hessabi, S. (2019). On the Scheduling of Energy-Aware Fault-Tolerant Mixed-Criticality Multicore *Systems with Service Guarantee Exploratio*n. IEEE Transactions on Parallel and Distributed Systems, 30(10), 2338–2354. doi:10.1109/TPDS.2019.2907846

Shooman, M. L. (2002). Reliability of computer systems and networks: Fault tolerance, analysis, and desig*n. New York: John Wiley & Sons Inc. doi:10.1002/047122460*X

Suen, T. T. Y., & Wong, J. S. K. (1992). Efficient task migration *algorithm for distributed systems. IEEE Transactions on Parallel and Distributed S*ystems, 3(4), 488–499. doi:10.1109/71.149966

Tosun, S., Kandemir, M., & Koc, H. (2006). Using task recomputation during application mapping in parallel embe*dded architectures. The International Conference on C*omputer Design, 6.

Tosun, S., Mansouri, N., Arvas, E., Kandemir, M., Xie, Y., & Hung, W. L. (2005). Reliability-centric hardware/software co-design. Proceedings of the *6th International Symposium on Quality of Electro*nic Design, 375-380. 10.1109/ISQED.2005.104

Xie, Y., & Hung, W. L. (2006). Temperature-aware task allocation and scheduling for e*mbedded multiprocessor systems-on-chip (mpsoc) design. J. of VLSI Sig. Proc.,* 177–189.

Xu, J., & Randell, B. (1996). Roll-Forward Error Recovery in Embedded Real-Time Systems. Int. Conf. on Parallel and Distr. Syst., 414-421.

Zhang, Y., Dick, R., & *Chakrabarty, K. (2*004). Energy-aware deterministic fault tolerance in distributed real-time embedded systems. Proceedi*ngs of the 41st annual Conference on De*sign Automation, 550-555. 10.1145/996566.996719

Zhu, D., Melhem, R., & Mosse, D. (2004). The Effects of Energy Management on Reliability in R*eal-Time Embedded Systems. International Conference on ICCAD,* 35-40.

Ziegler, J. F., Muhlfeld, H. P., Montrose, C. J., Curtis, H. W., O'Gorman, T. J., & Ross, J. M. (1996). Accelerated testing for cosmic soft-err*or rate. IBM Journal of Research* and Development, 40(1), 51–72. doi:10.1147/rd.401.0051

Chapter 5

Soft Computing Methods for Measuring Sustainability in the Agriculture Sector:
ISM Method to Develop Barriers of Agri-Sustainability in India

Suchismita Satapathy
https://orcid.org/0000-0002-4805-1793
KIIT University, India

ABSTRACT

Agriculture plays a vital role in the development of the Indian economy, and in addition, it contributes around 15% to the nation's GDP. Manually- or mechanically-operated diverse devices and supplies implied for farming machines are utilized in farming process. Still, sustainability is the most important issue in farming. Modern equipment smoke, dust, chemicals, and fertilizers both in manual-driven farming and modern farming are major environmental issues. So, in this chapter, sustainability issues in farming are studied, and a linear relationship between them can be found by interpretive structural modelling, such that the Micmac analysis and model can be developed for barriers of agricultural sector sustainability.

DOI: 10.4018/978-1-7998-1718-5.ch005

INTRODUCTION

Indian Economy depends on the Indian agriculture sector due to high investments for agricultural facilities, warehousing, and cold storage. Genetically modified crops and Organic farming has improved the fertility of land and crop production rate of Indian farmers. But the small and medium agricultural sector are following the traditional method of crop production as unable to purchase high-cost equipment and the conventional method of farming gives them many physical problems like lungs problem due to exposure to dust, and musculoskeletal disorders. Extreme weather conditions, the heavy workload during their working procedure gives them early old age, bone and muscle problems. So to attain better efficiency of performance and to improve the productivity of the worldwide farmers in the agricultural sector it is essential to design the tools and equipment keeping in consideration the farmer's capabilities and limitations. The tools and equipment design should be able to provide more human comfort, of good quality, more output focused and reduce the musculoskeletal injury. Occupational safety is a big issue of discussion for Agricultural workers. The method of working in field in extreme climate(heat,rain), contact with the chemicals(pesticides,fertilizers),the exposure to soil,dust,the contamination due to bacteria,exposure to animals,cattle, injury due to handtools and muskulateral disorders are the most important injuries faced by all Agriworkers. Agricultural workers need sufficient precaution and safety measures at the time of field and machine work, such that no physical damage occurs to them. Most of the Agricultural injuries result from the improper selection and use of hand tools. In the agricultural sector, traditional hand tools play a major role in performing farming activities. The conventional hand tools like spade/hoe, sickle, hammer, shovel, knife, etc. have been used since the ancient though some modifications are found now a day. As most of the farmers in India are from a poor economic background, they usually prefer conventional methods in farming instead of using the developed power operated machinery. The hand tools are mostly used in all farming activities like land preparation, weeding, harvesting of crops, etc. But tractors and other machineries are definitely solved injury and safety problems compared to the conventional tool. Modern equipment is not sustainable due to high noise, vibration, and pollution. Farming equipment modification, system design is essential to provide a better life to farmers but without environmental protection, social and economic stability is having no meaning. When all over the world is concerned about pollution farming policies must be framed to avoid pollution with improving productivity. As for safety and sustainability like two sides of a coin, both parts are important for the farming sector. So in this paper, an effort is taken to find a linear relationship between the farming processes by focusing sustainability issue, then a comparison between postures are done in conventional and modern farming practices.

BACKGROUND

Adarsh Kumar et al. (2008) have found 576 agricultural-related injuries with 332 i.e. 58% hand tool-related, from 9 villages with 19,723 persons in the 1st phase. Further, in the 2nd phase with more 21 villages of 78,890 persons, it was reported of 54 i.e.19% of hand tool-related out of 282 injuries. It was also recommended to have intervention development and training at block levels about the safety measures of equipment. Prasanna Kumar et al. (2009) have investigated the agricultural accident for six years i.e. between years 2000-2005 of 42 villages of 4 districts in Arunachal Pradesh in India. It was reported to have the accident rate as 6.39 per thousand workers per year with 40% farm implement related injury. S.K. Patel et al. (2010) have reported the agricultural accident rate as 0.8 per 1000 workers per year in Etawah district of Uttar Pradesh in India. Also, it is reported of a lack of study in agricultural injuries in developing countries due to the no availability of compiled information. Nilsson et al. (2010) have analyzed the responses from 223 injured farmers collected by the Swedish Farm Registry as part of a survey sent to 7,000 farms by the Swedish University of Agricultural Sciences and Statistics Sweden in 2004. These data showed that there were no significant differences in injuries incurred between the age groups, but that senior farmers seemed to suffer longer from their injuries. This study highlighted the importance of advising senior farmers to bear in mind that their bodies are no longer as young and strong as before. All age groups of farmers should, of course, be careful and consider the risks involved in their work, but since aging bodies need longer to heal, senior farmers probably need to be even more careful and review their work situation and work environment in order to avoid injuries during the final years of their working life. Then, It is recommended to educate senior farmers about the risks of injuries causing increasing damage due to their age. Cem Copuroglu et al. (2012) have studied the agricultural injuries of 41 patients for a period of 3 years and found hand as the most commonly injured part compared to the lower limb and foot. Roberto's Koekoeh K. Wibowo et al. (2016) have studied the farmers of East Java and Indonesia. Most of the farmer's injury was found in hand and the fatigue was reported as 92.8% in the upper back, 93.6% in mid-back and 91.8% in lower back. Safe, good and fit in hand were suggested as design criteria for hand tools. The recommended tool handles length was 12.4 cm and the diameter was 3cm, respectively. Many types of research are found on injury and equipment but very less research is found on Sustainability in the Agri industry. Mark(2008) has explored that factory farming's method of crowding and confining animals in warehouse-like conditions before killing them and mass-producing both "meat" from cows, pigs and chickens as well as dairy and eggs poses "an unacceptable level of risk to public health and damage to the environment. Mark (2008) has explored that tens of millions of tons of animal waste and agricultural chemicals into the environment every year—driving land, water and air pollution in the process:factory farms typically concentrate tens or hundreds of thousands of animals in one area, and a large operation can produce as much excrement as a small city. Heather(2007) has used EPA statistics, that means California's 1.8 million dairy cows produce as much waste as 36–72 million people. So, when taking into consideration tens or hundreds of thousands of animals, it's not surprising that this amounts to about 130 times more excrement than is produced by the entire human population every year. For many years, farmers have used animal manure to fertilize their fields, but factory farms produce far more waste than the land can absorb, turning disposal of this toxic by-product into a big problem for both the agriculture industry and society.

Wiley et.al (2004) have explained that like human waste, animal excrement from factory farms is not processed as sewage, making it about 500 times more concentrated than treated human waste while leaving pathogens and volatile chemicals intact. Farmers typically spray some liquidized manure onto the food being grown for animals using giant sprinkler jets, and store the rest in open-air cesspools that can be as large as several football fields and hold millions of gallons of waste. However, neither of these dispersal techniques is environmentally safe or sustainable.

The United Nations: Food and Agriculture Organization. have reported that of all the agricultural chemicals applied in the U.S., about 37 percent were used to grow crops for animals raised for food. Agricultural chemicals (or agrichemicals) refer to the wide variety of chemical products used in agriculture, such as pesticides (including insecticides, herbicides and fungicides), as well as synthetic fertilizers, hormones, and antibiotics. Farmers spray agricultural chemicals onto food is grown for animals in order to kill bugs, rodents, weeds, and other organisms that would otherwise supplant or eat the grain grown for the animals. They also apply these substances directly to animals' skin, fur or feathers to combat insect infestation.

RESEARCH METHODOLOGY

A total of 145 farmers from five villages were selected randomly from the Odisha state in India. Prior consent was obtained from the respective village heads or leaders. The author and co-author visited the farmers personally and collected the information and data collected using a Likert scale and by standardized nordic questionnaires as shown in Appendix 1. Then, seven experts from different areas of expertise were consulted for subsequent decision analysis.

Through extensive literature review and using expert opinion different processes involved in rice farming, and different farming pollution-related which are disorders and discomforts were identified as Different processes in farming.

- Land preparation process (LPP)
- Seed soaking and scattering/Fertilizer applying process (SFP)
- Pesticides mixing and spraying process (PMSP)
- Weed pulling process (WPP)
- Rice harvesting process (RHP)
- Threshing process (TP)

The Questionnaire designed to find pollution occurs in the above processes of rice farming that produces air, water and land pollution. Then 145 data is collected by using Likert scale 1 to 5(1 = totally disagree, 2 = partially disagree, 3 =No opinion, 4 =Partiallyagree, 5 = totally agree) and analyzed it is found that fertilizers and chemicals produce maximum pollution. Then the farming processes are ranked by Prometheus-2, to select most polluting process, such that the policies must be framed by stakeholders to avoid it and make the sustainable farming process.

Table 1 shows Barriers of sustainable farming process in Indian rice farming.

RESULT AND DISCUSSION

ISM developed by Warfield is a qualitative tool has the ability to decompose a complicated system into several subsystems and construct a multilevel structural model utilizing the practical experience and knowledge of a group of experts(Kumar et al., 2013). Owing to its ability to develop a map of the complex relationships between the various elements involved in a complex situation, it is often used to provide fundamental understanding of complex situations, as well as to put together a course of action for solving a problem(Ansari et al., 2013). ISM is a combination of three modeling languages viz: words, digraphs and discrete mathematics, to offer a methodology for structuring complex issues(Ansari et al., 2013).The reason behind choosing this methodology are as follows:• It allows in the identification of the main research variables, and also it can act as a tool for imposing order and direction on the the complexity of relationships among variables (Kannan and Haq 2007; Govindan et al., 2015).• It can extract the specific relationships among the variables and portray them in a graphical model(Borade and Bansod, 2012)ISM is powered by two important concepts transitivity and reachability which makes the ISM based model simple and understandable. The two concepts can be can be explained as follows: let there be three factors "i", "j" and "k" where the variable "j" is driven by the variable "i" directly and the variable "k" is driven by "j" as well as "i" then the relationship between "i" and "j" as well as "j" and "k" can be shown by two arrows with the arrowhead pointing towards "j" and "k" respectively. In this case the relationship between "i" and "j" need not be shown using an arrow. The

Table 1. Barriers of Sustainable Farming process

No.	Farming Process and Symptoms	Number of Farmers (%)
Land Preparation Process		
1.	Irritation from smoke during burning rice stalks, if any.	05 (3.47)
2.	Exposure to diesel fumes from plowing and harrowing tractors.	08 (5.55)
3.	Exposure to loud noise and vibration from plowing and harrowing tractors.	08 (5.55)
4.	Dust spreads due to tractor	08(5.20)
Seed Soaking and Scattering/Fertilizer Applying Process		
5.	Exposure to wet and humid soil.	28 (19.44)
Pesticides Mixing and Spraying Process		
6.	Exposure to vibration of pesticide applying machines.	02 (1.38)
7.	chemical pollution from pesticide exposures:	0.44(22.28)
8.	Fertilizers create air pollution	0.48(27.8)
9.	Animal waste also creates air pollution.	0.46(23.9)
	Rice harvesting process	22 (15.27)
	Threshing process	23 (15.97)
10.	Exposure to vibration and noise of threshing machines.	20 (13.88)
11.	Exposure to grain dusts.	18(20.3)

reachability grid helps in making the different levels in ISM. The different strides required in the ISM strategy are given underneath, furthermore appeared in figure 1.)

Steps for constructing ISM based model are as follows:

3.1 Structural Self-Interaction Matrix (SSIM)

Contextual relationships, existing among the enablers were identified with consultation with a group of experts comprising of academicians and the industry people in this research. In order to analyze the relationship among the SSCM enablers, a contextual relationship of 'leads to' type is chosen. Keeping in mind the the contextual relationship for each variable, the existence of a relation between any two variables (i and j) and the associated direction of the relation is questioned. Four symbols are used to denote the direction of relationship between the variables (i and j):

Figure 1. Flow diagram for ISM construction

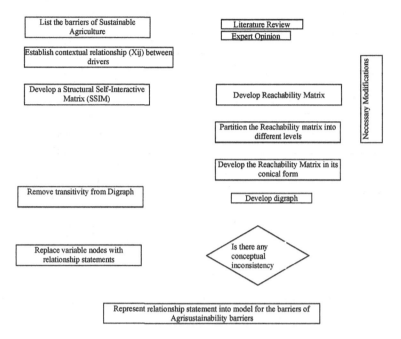

V: Variable *I* will help to alleviate variable *j*

A: Variable *j* will be alleviated by variable *i*

X: Variable *i* and *j* will help to alleviate each other

O: Variable *i* and *j* are unrelated.

Based on the contextual relationships, the SSIM is developed for the 11 barriers (Table 2 shows Structural Self-Interaction Matrix (SSIM)).

3.2 Reach Ability Matrix

The SSIM is transformed into a binary matrix, called the initial reach ability matrix by substituting V, A, X and O by 1 and 0 as per the case. The rules for the substitution of 1s and 0s are as follows:

- if the (i, j) entry in the SSIM is V, then the (i, j) entry in the reach ability matrix becomes 1 and the (j, i) entry becomes 0
- if the (i, j) entry in the SSIM is A, then the (i, j) entry in the reach ability matrix becomes 0 and the (j, i) entry becomes 1
- if the (i, j) entry in the SSIM is X, then the (i, j) entry in the reach ability matrix becomes 1 and the (j, i) entry becomes 1

Table 2. Structural Self-Interaction Matrix (SSIM)

Barriers	11	10	9	8	7	6	5	4	3	2	1
1.	V	V	A	A	A	V	V	A	A	A	X
2.	V	V	V	V	X	V	V	X	V	X	V
3.	V	V	X	X	A	V	V	A	X	A	V
4.	V	V	V	V	X	V	V	X	V	X	V
5.	A	V	A	A	A	V	X	A	A	A	A
6.	A	V	A	A	A	X	A	A	A	A	A
7.	V	V	V	V	X	V	V	X	V	X	V
8.	V	V	X	X	A	V	V	A	X	A	V
9.	V	V	X	X	A	V	V	A	X	A	V
10.	A	X	A	A	A	A	A	A	A	A	A
11.	X	V	A	A	A	V	V	A	A	A	A

- if the (i, j) entry in the SSIM is O, then the (i, j) entry in the reach ability matrix becomes 0 and the (j, i) entry becomes 0.

Following these rules, initial reach ability matrix for the sustainable barriers in Agrifarming of India is identified and the final reach ability matrix is obtained by incorporating the transitivity's, this is shown in Table III. Transitivity that helps in maintaining the conceptual consistency in ISM methodology, implies that if an element 'k' relates to an element 'j' and the element 'j' to element 'i', then the element 'k' relates to element 'i'. Table 3 shows the Reach ability Matrix.

3.3 Level Partitions

The reachability and antecedent set for each variable are obtained from the final reach ability matrix. The reach ability set for a particular variable consists of the variable itself and the other variables, which may help to achieve. The antecedent set consists of the variable itself and the other variables, which may help in achieving them. Subsequently, the intersection of these sets is derived for all variables. The variable for which the reach ability and the intersection sets are the same is assigned as the top-level variable in the ISM hierarchy as it would not help to achieve any other variable above their own level. After the identification of the top-level element, it is discarded from the list of remaining variables. It can be observed From Table IV, that 'Exposure to vibration and noise of threshing machines.' (variable 10) is found at level I. Thus, it would be positioned at the top of the ISM hierarchy. This

Table 3. Reachability Matrix

Enablers	11	10	9	8	7	6	5	4	3	2	1	Driving Power
1.	1	1	0	0	0	1	1	0	0	0	1	5
2.	1	1	1	1	1	1	1	1	1	1	1	11
3.	1	1	1	1	0	1	1	0	1	0	1	8
4.	1	1	1	1	1	1	1	1	1	1	1	11
5.	0	1	0	0	0	1	1	0	0	0	0	3
6.	0	1	0	0	0	1	0	0	0	0	0	2
7.	1	1	1	1	1	1	1	1	1	1	1	11
8.	1	1	1	1	0	1	1	0	1	0	1	8
9.	1	1	1	1	0	1	1	0	1	0	1	8
10.	0	1	0	0	0	0	0	0	0	0	0	1
11.	1	1	0	0	0	1	1	0	0	0	0	4
Dependence Power	8	11	6	6	3	10	9	3	6	3	7	

iteration is repeated until the levels of each variable are found out. The identified levels aids in building the digraph and the final model of ISM (Table 4).

Table 4 shows level partition.

3.4 Formation of ISM-based Model

The structural model is generated from the final reach ability matrix and the digraph is drawn. Removing the transitivity as described in the ISM methodology, the digraph is finally converted into the ISM as shown in Figure 3.

3.5 MICMAC Analysis

The MICMAC (Matrice d'Impacts Croisés Multiplication Appliquée á un Classement (cross-impact matrix multiplication applied to classification)) analysis is used to identify and to analyze the driving power and the dependence of the selected issues. The MICMAC rule, in view of the duplication properties of lattices, expresses that if issue A specifically impacts issue B and B straightforwardly impacts issue C, then any change influencing A can have repercussions on C (Sharma et al 1995). In this study, the MICMAC analysis is used to analyze the driving and dependent power of selected enablers in the SSCM connection. The driving power and dependence of each of these enablers are calculated in Table III. In table III, an entry of "1" along the column and rows indicates the dependence and driving power, respectively.

Table 4. Level partition Iteration 1-5

Sl/No	Factors	Reachability Set	Antecedent Set	Intersection	Level
1	Irritation from smoke during burning rice stalks, if any.	1,5,6,10,11	1,2,3,4,7,8,9	1	V
2	Exposure to diesel fumes from plowing and harrowing tractors.	1,2,3,4,5,6,7,8,9,10,11	2,4,7	2	VII
3	Exposure to loud noise and vibration from plowing and harrowing tractors.	1,3,5,6,8,9,10,11	2,3,4,7,8,9	3	VI
4	Dust spreads due to tractor	1,2,3,4,5,6,7,8,9,10,11	2,4,7	4	VII
5	Exposure to wet and humid soil	5,6,10	1,2,3,4,5,7,8,9,11	5	III
6	Exposure to vibration of pesticide applying machines.	6,10	1,2,3,4,5,6,7,8,9,11	6	II
7	chemical pollution from pesticide exposures:	1,2,3,4,5,6,7,8,9,10,11	2,4,7	7	VII
8	Fertilizers create air pollution	1,3,5,6,8,9,10,11	2,3,4,7,8,9	8	VI
9	Animal waste also creates air pollution.	1,3,5,6,8,9,10,11	2,3,4,7,8,9	9	VI
10	Exposure to vibration and noise of threshing machines.	10	1,2,3,4,5,6,7,8,9,10,11	10	I
11	Exposure to grain dusts.	5,6,10,11	1,2,3,4,7,8,9,11	11	IV

Further, the Sustainable Agri barriers are categorized into four clusters according to their driving and dependence power.

In the MICMAC examination, the chosen issues are arranged into four clusters (Figure 2). The primary group (I) contains "self-ruling variables" that have weaker driver power and less dependence. These boundaries are generally disengaged from the framework, with which they have just a few connections, which might be solid. Second cluster (II) contains "subordinate variables" that have less driver control however solid dependence. The third group (III) has the linkage variables that have strong driving power and strong dependence. These SSCM enablers are shaky in the way that any activity on these enablers will affect others furthermore an input on themselves. Fourth cluster (IV) incorporates the autonomous variables having solid driving power yet frail reliance. Thusly, the driver power-reliance outline is built which is appeared in Figure 5. As delineation, it is seen from Table III that barrier 1(Irritation from smoke during burning rice stalk is having a driving force of "5" and dependence of "7". Accordingly, in this figure, it is situated at a spot relating to a driving force of "5" and a reliance of "7" i.e. cluster IV.

CONCLUSION

Sustainability is a very critical issue in farming. If farming is traditional, manures of animal waste are used instead of chemicals and fertilizers to improve farming productivity. But the use of the wastes of animals also creates pollution. Two of the many possible practices of sustainable agriculture are crop rotation and soil amendment, both designed to ensure that crops being cultivated can obtain the necessary nutrients for healthy growth. Soil amendments would include using locally available compost from community recycling centers. These community recycling centers help produce the compost needed by the local organic farms. *Sustainable agriculture* is a type of *agriculture* that focuses on producing long-term crops and livestock while having minimal effects on the environment. This type of *agriculture* tries to find a good balance between the need for food production and the preservation of the ecological system within the environment.

Figure 2. Driver and Dependence Power Diagram

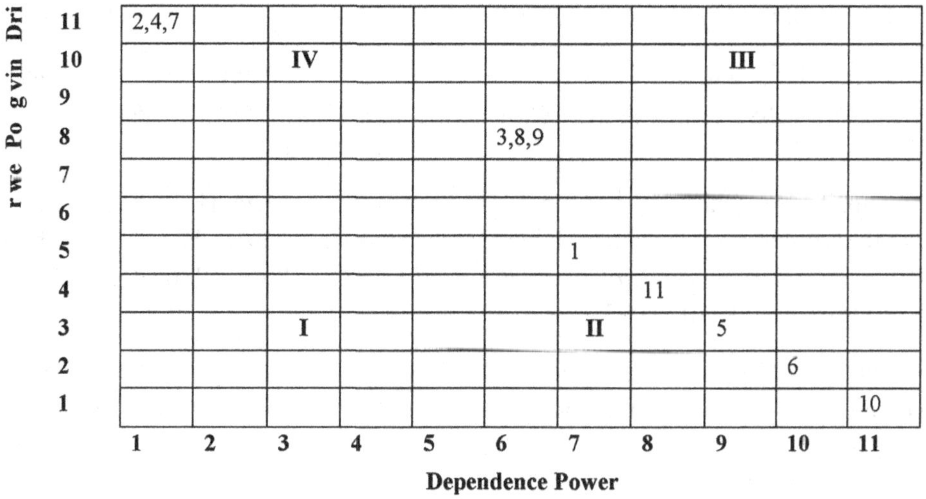

		1	2	3	4	5	6	7	8	9	10	11
	11	2,4,7										
	10			IV						III		
	9											
	8						3,8,9					
	7											
	6											
	5							1				
	4								11			
	3			I				II		5		
	2										6	
	1											10

Dependence Power

REFERENCES

Agarwal, A., Shankar, R., & Tiwari, M. K. (2006). Modeling agility of supply chain. *Industrial Marketing Management, 36*(4), 443–457. doi:10.1016/j.indmarman.2005.12.004

Banthia, J. K. (2004). *Census of India 2001 - Primary Census Abstracts.* Registrar General & Census Commissioner, Govt. of India.

Cooper, S. P., Burau, K. E., Frankowski, R., Shipp, E., Deljunco, D., Whitworth, R., ... Hanis, C. (2006). A cohort study of injuries in migrant farm worker families in South Texas. *Annals of Epidemiology, 16*(4), 313–320. doi:10.1016/j.annepidem.2005.04.004 PMID:15994097

DeMuri, G. P., & Purschwitz, M. A. (2000). Fatal injuries in children: A review. *WMJ: Official Publication of the State Medical Society of Wisconsin, 99*(9), 51–55. PMID:11220197

Gite, L. P., & Kot, L. S. (2003). *Accidents in Indian Agriculture.* Technical Bulletin No. CIAE/2003/103. Coordinating Cell, All India Coordinated Research Project on Human Engineering and Safety in Agriculture, Central Institute of Agricultural Engineering, Nabibagh, Bhopal.

GOI. (2002). *India Vision 2020. Planning commission.* New Delhi: Govt. of India.

GOI. (2006). *Population projections for India and States 2001-2006, Report of the technical group on population projections constituted by the National Commission on Population.* New Delhi: Govt. of India.

Helkamp, J., & Lundstrom, W. (2002). Tractor Related Deaths Among West Virginia Farmers. *Annals of Epidemiology, 12*(7), 510. doi:10.1016/S1047-2797(02)00344-7

Hope, A., Kelleher, C., Holmes, L., & Hennessy, T. (1999). Health and safety practices among farmers and other workers: A needs assessment. *Occupational Medicine (Philadelphia, Pa.), 49*, 231–235. PMID:10474914

Huiyun, X., Zengzhen, W., Lorann, S., Thomas, J. K., Xuzhen, H., & Xianghua, F. (2000). Agricultural work-related injuries among farmers in Hubei, People's Republic of China. *American Journal of Public Health, 90*(8), 1269–1276. doi:10.2105/AJPH.90.8.1269 PMID:10937008

Karthikeyan, C., Veeraragavathatham, D., Karpagam, D., & Ayisha, F. S. (2009). Traditional tools in agricultural practices. *Indian Journal of Traditional Knowledge, 8*(2), 212–217.

Knapp, L. W. Jr. (1965). Agricultural Injuries Prevention. *Journal of Occupational Medicine.*, *7*(11), 553–745. doi:10.1097/00043764-196511000-00001 PMID:5831719

Knapp, L. W. (1966). Occupational and Rural Accidents. *Archives of Environmental Health*, *13*(4), 501–506. doi:10.1080/00039896.1966.10664604 PMID:5922020

Kumar, A., Singh, J. K., Mohan, D., & Varghese, M. (2008). Farm hand tools injuries: A case study from northern India. *Safety Science*, *46*(1), 54–65. doi:10.1016/j.ssci.2007.03.003

Kumar, A., Varghese, M., Pradhan, C. K., Goswami, A., Ghosh, S. N., & Nag, P. K. (1986). Evaluation of working with spade in agriculture. *The Indian Journal of Medical Research*, *84*, 424–429. PMID:3781600

Lombardi, M., & Fargnoli, M. (2018). Prioritization of hazards by means of a QFD-based procedure. *International Journal of Safety and Security Engineering.*, *8*(2), 342–353. doi:10.2495/SAFE-V8-N2-342-353

Mishra, D., & Satapathy, S. (2018). Drudgery Reduction of Farm Women of Odisha by Improved Serrated Sickle. *International Journal of Mechanical Engineering and Technology*, *9*(2), 53–61.

Mittal, V. K., Bhatia, B. S., & Ahuja, S. S. (1996). *A Study of the Magnitude, Causes, and Profile of Victims of Injuries with Selected Farm Machines in Punjab. Final Report of ICAR adhoc Research Project*. Ludhiana, India: Department of Farm Machinery and Power Engineering, Punjab Agricultural University.

Mohan, D. (2000). Equipment-related injuries in agriculture: An international perspective. *Injury Control and Safety Promotion*, *7*(3), 1–12.

Mohan, D., & Patel, R. (1992). Design of safer agricultural equipment: Application of ergonomics and epidemiology. *International Journal of Industrial Ergonomics*, *10*(4), 301–309. doi:10.1016/0169-8141(92)90097-J

Moore, H. (2007). You Can't Be a Meat-Eating Environmentalist. *American Chronicle*. Retrieved from http://www.americanchronicle.com/articles/view/24825

Murphy, D. J. (1992). *Safety and Health for Production Agriculture*. St. Joseph, MI: American Society of Agricultural Engineers.

Nag, P. K., & Nag, A. (2004). Drudgery, accidents and injuries in Indian agriculture. *Industrial Health*, *42*(4), 149–162. doi:10.2486/indhealth.42.149 PMID:15128164

Nilsson, K., Pinzke, S., & Lundqvist, P. (2010). Occupational Injuries to Senior Farmers in Sweden. *Journal of Agricultural Safety and Health, 16*(1), 19–29. doi:10.13031/2013.29246 PMID:20222268

Patel, S. K., Varma, M. R., & Kumar, A. (2010). Agricultural injuries in Etawah district of Uttar Pradesh in India. *Safety Science, 48*(2), 222–229. doi:10.1016/j. ssci.2009.08.003

Prasanna Kumar, G. V., & Dewangan, K. N. (2009). Article. *Safety Science, 47*, 199–205. doi:10.1016/j.ssci.2008.03.007

Raj, T., & Attri, R. (2011). Identification and modeling of barriers in the implementation of TQM. *International Journal of Productivity and Quality Management., 28*(2), 153–179. doi:10.1504/IJPQM.2011.041844

Rautiainen, R. H., & Reynolds, S. J. (2002). Mortality and morbidity in agriculture in the United States. *Journal of Agricultural Safety and Health, 8*(3), 259–276. doi:10.13031/2013.9054 PMID:12363178

Ravi, V., & Shankar, R. (2005). Analysis of interactions among the barriers of reverse logistics. *Technological Forecasting and Social Change, 72*(8), 1011–1029. doi:10.1016/j.techfore.2004.07.002

Robertoes Koekoeh, K. W., & Peeyush, S. (2016). Article. *Agriculture and Agricultural Science Procedia., 9*, 323–332. doi:10.1016/j.aaspro.2016.02.142

Voaklander, K. D., Kelly, D. C., Rowe, B. H., Schopflocher, D. P., Svenson, L., Yiannakoulias, N., & Pickett, W. (2006). Pain, medication, and injury in older farmers. *American Journal of Industrial Medicine, 49*(5), 374–382. doi:10.1002/ ajim.20292 PMID:16526061

Wiley, K., Vucinich, N., Miller, J., & Vanzi, M. (2004, November). *Confined Animal Facilities in California*. Retrieved from http://sor.senate.ca.gov/sites/sor.senate. ca.gov/files/%7BD51D1D55-1B1F-4268-80CC-C636EE939A06%7D.pdf

Zhou, C., & Roseman, J. M. (1994). Agricultural injuries among a population based sample of farm operators in Alabama. *American Journal of Industrial Medicine, 25*(3), 385–402. doi:10.1002/ajim.4700250307 PMID:8160657

KEY TERMS AND DEFINITIONS

Manual-Driven Farming: Hand tools used for farming.

Modern Farming: Machines like tractors and threshers used in farming.

Sustainability: Sustainable agriculture is a type of agriculture that focuses on producing long-term crops and livestock while having minimal effects on the environment.

Chapter 6

A Hybrid Analysis Method for Train Driver Errors in Railway Systems

Mohammed Manar Rekabi
Norwegian University of Science and Technology, Norway

Yiliu Liu
Norwegian University of Science and Technology, Norway

ABSTRACT

The main objective of this chapter is to analyze safety in railway systems through studying and understanding the train drivers' tasks and their common errors. Different approaches to classifying and analyzing driver errors are reviewed, as well the factors that affect driver performance. A comprehensive overview of the systems theoretic process analysis (STPA) method is presented, along with how it could be applied for controllers and humans. Quantitative risk assessment, along with some methods for quantifying human errors, are overviewed, and a Bayesian network is selected to study the effects of the identified driver errors. A case study aims to present a detailed quantitative safety analysis at European Train Control System (ETCS) system Levels 1 and Level 2, including driver errors. The STPA and Bayesian methods are combined to identify the hazards and quantify the probabilities of hazards when trains fail to stop at red signals.

DOI: 10.4018/978-1-7998-1718-5.ch006

INTRODUCTION

The introduction of new technologies and digitalized solutions in railway systems has led to increased complexity, and thus to the emergence of unintended system performance. Train drivers are also changing their roles to the supervisors of more automatic systems.

Many investigations have been conducted to identify vulnerabilities in railway systems and reinforce railway safety (Baysari, Caponecchia, McIntosh, & Wilson, 2009; Felipe, Sallak, Schön, & Belmonte, 2013), and human errors are found as the most significant source of accidents or incidents in railway systems (Felipe et al., 2013; Kyriakidis, Pak, & Majumdar, 2015). For example, at least 75% of fatal accidents in European railway systems between 1990 and 2013 occur due to human errors. In UK, accidents due to driver error made up more than 50% of all accidents between 1945 and 2012 (Kyriakidis et al., 2015). The ratio of accidents associated with human errors increases to more than 80% of all major railway accidents worldwide (Kyriakidis, Majumdar, & Ochieng, 2018).

Drivers need to integrate the various sources of information to achieve the following goals: 1) moving the train according to the authority of movement, 2) moving the train within the safe speed limits, and 3) making safe and accurate scheduled stops (Hamilton & Clarke, 2005). It is obvious that any errors of drivers will be dangerous and can result in fatal accidents if no effective further mitigation measures work.

Queensland Rail Driver Training Centre have identified three major types of driver errors: wrong brakes (too early or too late), failure to respond to in-cab station protection and vigilance systems and driving at a speed that exceeds the limit (Dorrian, Roach, Fletcher, & Dawson, 2007). Australian Transport Safety Bureau (ATSB) (Baysari et al., 2009) highlights that major driver errors occur when drivers fail to detect and respond to signals or when they cannot judge aspects of the train correctly. The impact of errors depends on both the speed of error detection and the effectiveness of recovery.

The underlying factors causing driver errors are summarized in (Cacciabue, 2005): inadequate communication with crew, poor interaction with equipment, weak compliance with rules, and lack knowledge and experience. In addition, Hamilton & Clarke (2005) highlight that driver errors occur when the driver meets limited time for executing a task or limited resources within the cognitive system. The driver performance depends on how and when the driver controls speed, observes signs, signals and other visual targets, responds to safety devices, and acts with requirements of stations.

The Human Factors Analysis and Classification System (HFACS) released by ATBS (Baysari et al., 2009), divides drive failures into: failure of organizational influences, supervisory factors, preconditions for unsafe acts, and unsafe acts.

However, no current literature has directly linked errors of train drivers with quantitative safety analysis, although the latter is regarded as one of the most important techniques in ensuring railway safety.

On the other hand, variety of equipment and technology are coming into railway systems and are influencing driver behaviors. For instance, in the new European Rail Traffic Management System (ERTMS), more automatic train protection (ATP) systems are introduced to stop trains when necessary without any actions from the drivers (Castillo et al., 2016), and automatic warning systems (AWS) give drivers an audible and visual indication of the status of the signal ahead (McLeod, Walker, & Moray, 2005). Train drivers need to obtain data by monitoring a specific cab, make decisions according to the data, and enable functions via buttons on the cab (Ghazel, 2014). The normal operation of such a process is highly dependent on software programs, and there are significant human-machine interactions (Callet, Fassi, Fedeler, Ledoux, & Navarro, 2014). New types of hazards generate, and the investigation of unsafe actions within these interactions and dependencies calls for new initiatives of hazard identification.

In this paper, we will adopt a new approach to reveal the hazards related to train drivers, based on the understanding of driver duties and the review the existing hazard identification methods. For those main identified hazards, we will explore a suitable approach for further quantitative safety assessment, which can be more informative to decision-makers.

The remainder of this paper is as following: Section 2 briefly introduces the systems theoretic process analysis (STPA) for hazard identification, and section 3 presents the Bayesian network approach for quantitative analysis. The two methods will be combined in section 4 for studying the effects of driver errors on two railway control systems – ETCS Level 1 and Level 2.

HAZARD IDENTIFICATION FOR TRAIN DRIVERS

Methodology of Hazard Identification

A hazard is a source of danger that may cause harm to an asset (Rausand, 2011). Hazard identification, in railway, means to reveal all possible sources where assets around a railway system, e.g. trains and passengers may be harmed by the actions of a train driver.

As mentioned above, the hazards to the assets around a railway system can come from human-machine interactions, as well as other social and organizational factors, rather than the unreliability of components. A more thorough hazard identification should be dependent on a model of accident causality for understanding scenarios of accidents and causal factors.

Many accident causality models are existing in the fields of risk assessment. For example, the energy and barrier models explain the occurrence of an accident as the failure of barriers to separate uncontrolled energies from vulnerable targets (Rausand, 2011). The Domino model describes accident causation as a chain of discrete events in a particularly temporal order, and an accident in the Swiss cheese model is the result of many factors (manifest and latent) combined in unsafe ways (Rausand, 2011). These accident causality models are very informative in some cases, but they do not take dependencies between barriers into consideration, which may lead to the arbitrary conclusions, where equipment vendors or drivers are incorrectly blamed alone for an accident. In addition, these models treat humans in a very simplified way, without considering the environmental influences. For example, in the context of modern railway systems, the following questions need to be answered when we want to examine the risks of accidents: Is the driver able to interpret data correctly? Does the driver understand the system and know how it will behave with some control actions?

In this paper, we adopt the Systems-Theoretic Accident Model and Processes (STAMP) to investigate hazards related with train drivers. Such an approach is based on control theory, treats the system as a set of dynamic processes, and describe the system performance as a whole rather than in terms of cause-effect mechanisms (Rausand, 2011). In STAMP, safety is realized by controlling interactions among the system components and enforcing safety constraints rather than preventing failures of components. Hence, unsafety is a control problem rather than a reliability problem, while an accident occurs in the case of inadequate controls (Leveson, 2012). Inadequate controls are due to different reasons, e.g. missing safety constraints, inadequate commands for safety control, and inadequate communication (Leveson, 2012). Humans are also possible to be involved in the analysis of STAMP.

Systems theoretic process analysis (STPA) is under the framework of STAMP, with the aim to identify the potential hazards that can lead to accidents (Leveson, 2012). STPA depends on a functional control diagram rather than a physical component diagram, and thus such a method can be conceived of as a top-down hazard analysis technique that includes a set of steps from the basic diagram to the determination of the way control actions might provoke hazards (Plioutsias & Karanikas, 2015). More information about STPA can be found in (Leveson, 2012).

Hazard identification with STPA

In general, STPA includes 5 steps: Identification of hazardous events, identification of safety constraints, building of the hierarchical safety control structure, identification of unsafe control actions, and identification of causal scenarios.

The behaviors of a train driver are treated in this paper by using an additional process model in STPA. Driver errors are considered as the result of incorrect behaviors, which is modelled in terms of conditions and objectives of the decision-maker. The thinking and motivations of a driver are basic components of generating causal analysis scenarios in STPA (Leveson, 2012). Drivers employ dynamic control actions within a loop of feedback (e.g. weather conditions and current speed) and goals (e.g. obeying the speed limit or arriving at the destination at a required time), as shown in Fig. 1.

Most of the work of a train driver is kind of a decision-making. In case of state changes, the driver needs to take actions accordingly. To identify driver errors and the relevant factors, it is necessary to understand how the driver makes a decision, and how flaws in the decision-making occur.

The decision-making is based on the information that the driver obtains from displays, and lineside signs. Wrong decisions can come from the absence of the needed information, e.g. signaling failure, unnecessary or incorrect data, incorrect judgment on the actual state, and wrong beliefs about what the system can do and will do. The rightness of actions after decision-making also depends on many factors, such as time pressure, context, environment, and personalities (e.g. fatigue, stress, expectations, experience, and skills) (France, 2017).

Figure 1. An example of safety control loop

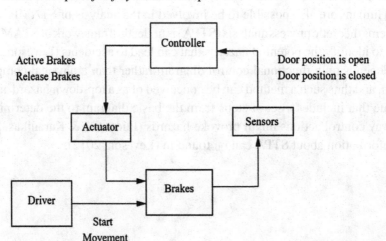

QUANTITATIVE RISK ASSESSMENT

After determining the hazards associated with train drivers and their causes, it is necessary to conduct quantitative analysis to estimate both the probability of the driver errors that lead to different accident scenarios, and the consequences of these accidents.

Performance Shaping Factors of Drivers

Performance-shaping factors (PSFs) have been used in several studies (see Wilson, Farrington-Darby, Cox, Bye, & Hockey, 2007; Felipe et al., 2013; Kyriakidis et al., 2018) for evaluating human performance. In the EPSRC Rail Research UK program (Wilson et al, 2007), the most important PSFs are identified as route knowledge (training - familiarity), in-cab environment (noisy - comfortable), workload (high - low), psychological components (e.g. vigilance - fatigue), and the procedures and violations. Felipe et al., (2013) highlight that the main PSFs for driver errors are human habitude, the physical condition of the driver, inefficient training or sensory defects, and the mental condition of the driver. Hamilton & Clarke (2005) list 43 factors in seven main categories, including: dynamic personal, personal, task, team, organizational, system and environmental factors. The results from these studies can be used to develop appropriate targeted mitigation measures.

Bayesian Network as the Quantitative Approach

There are numerous methods used in quantitative human reliability and error analysis. For instance, the technique of human error rate prediction (THERP) is one of widely used methods, which can calculate human error probability (HEP) within different tasks by a human, e.g. a train driver (Rausand, 2011). The human error assessment and reduction technique (HEART) determines HEP values by applying a set of nominal HEPs and weighting factors (Hamilton & Clarke, 2005). Another type of methods focus on the contextual conditions (human factors) under which a given action is performed (Marseguerra, Zio, & Librizzi, 2006). For example, the cognitive reliability and error analysis method (CREAM), calculates the probability of human error depending on the level of control. However, these methods neglect the interdependences among the PSFs, so that it is difficult to use them to quantify the human-machine interactions in railway systems.

An alternative approach is the Bayesian network (BN), which is characterized by its ability to model uncertainties and the dependencies, as well as the adaptability to different sources of data (e.g. expert judgements and historical data) (Lee & Song, 2016). In addition, since BN depends on a causal and graphical model with explanatory and predictive capability, it can not only predict the probability of hazardous events, but also explain why such hazardous events can happen (Mu, Xiao, Xue, & Yuan, 2015). Therefore, we select BN in this paper as the approach to quantify identified driver errors.

A BN is composed of a directed acyclic graph (DAG) and conditional probability tables (CPTs). The DAG means that cycles are not allowed in the network. A BN is represented by a finite set of nodes (variables that a states or conditions) and a set of directed arcs (denoting causal influence between nodes), while CPTs are a set of conditional probability tables that express the strength of the relationships between the variables (Hamilton & Clarke, 2005). More information about BN can be found in (Rausand, 2011; Castillo, Grande, Mora, Lo, & Xu, 2017).

The Bayesian network is considered a suitable approach to model human performance (for instance, train driver in a railway system), because it can (Mu et al, 2015):

1. Use multiple sources of data to estimate the values of PSFs;
2. Include different PSFs with various probability distributions (different numbers of states) in the model;
3. Take dependency among PSFs into consideration;
4. Determine which PSFs have a significant influence on the final risk.

In the next chapter, a hybrid method of STPA and Bayesian network will be implemented via a case study.

AN ILLUSTRATIVE EXAMPLE FOR THE HYBRID METHOD

Case Study

Train drivers safely pass tens of thousands of signals every year. Only on very rare occasions does a signal passed at danger (SPAD) event occur; if this happens, a serious accident (collision or derailment) with many fatalities could ensue. Field investigations and analysis of data from previous accidents show that most SPADs occur due to a combination of operational errors, environmental conditions, errors associated with human performance, and interactions between components themselves (Pasquini, Rizzo, & Save, 2004).

To eliminate the different signaling and control systems in different European countries, a recent European standard (ERTMS) has been developed with the aim to improve the safety, reliability, performance, and interoperability of the European rail network (Flammini, Marrone, & Mazzocca, 2006). ERTMS system consists of heterogeneous, distributed components that are classified into three groups: European Train Control System (ETCS), European Traffic Management Layer (ETML), and Global System for Mobile Communications for Railway (GSM-R).

According to many articles and authors, ETCS and GSM-R are considered the most important parts of ERTMS. Different equipment is used at each level and three main subsystems (Flammini, Marrone, Iacono, Mazzocca, & Vittorini, 2014): lineside-, onboard- and trackside subsystems. The ERTMS/ETCS specifications define four application levels of ERTMS/ETCS operation; the main difference between these levels is the manner of interaction between trackside equipment and the trains (McLeod et al., 2005). In this chapter, we will study ERTMS/ETCS Level 1 and Level 2, and to reveal and compare driver risks to the SPAD accidents of these two systems.

System Description

ERTMS/ETCS – Level 1&2

A signaling system designed to be compatible with conventional signaling systems and that aims to provide automatic train protection functions (ATP) by using Balises (Eurobalises) and transmission loops (Euroloops) (Barger, Schön, & Bouali, 2009). ERTMS/ETCS level 1 with its different components is shown in Fig 2(a). It can be observed that lineside signals and train detectors are maintained, while communication between trackside and train is ensured by Balises (Eurobalises), which are located along the track next to lineside signals at required distances and connected to the train control centre through the lineside electronic unit (LEU). Their role is to transmit track description data and movement authority to the onboard subsystem. The onboard computer (EVC) continuously monitors and calculates the maximum speed of the train and the braking curve, as well as determining the next braking point if needed. This is achieved by relying on train braking characteristics and the data sent by Balises (McLeod et al., 2005). This information is displayed on the MMI to make it available to the driver.

While for ERTMS/ETCS Level 2 (as shown in Fig. 2(b)), a digital radio-based signaling system functioning as a train protection system. Accordingly, all necessary data, such as speed profile and movement authority (which depends on track-specific data and the status of specific signals identified by using conventional track circuits on the route ahead) are transmitted to the EVC unit directly from an RBC via the GSM-R link. The Euroradio protocol is used to achieve this link. From the other side, the exact position, speed, and direction of the train is sent periodically to RBCs from the onboard system via GSM-R link (Flammini et al.,2014). ERTMS/ETCS level 2 is thus a highly sophisticated system that monitors the train's travel

Figure 2a. ERTMS/ETCS Level 1

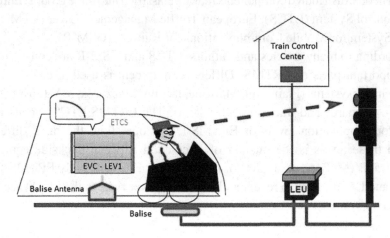

and advises the driver if he passes a red (danger) signal or exceeds a speed limit. In this system, lineside signals are optional and Balises are used only as reference points for correcting distance measurement errors; this is because their functions are limited to the transmission of static messages, such as location, track profile and speed limit, to the onboard subsystem (McLeod et al., 2005).

Figure 2b. ERTMS/ETCS Level 2

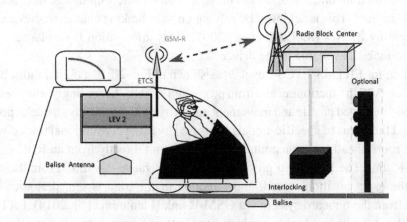

Signal Passed at Danger (SPAD)

Railway lines around the world are divided into sections called blocks, each of which is protected by a signal. Accordingly, permission to enter the next block can be given to the train if and only if there is no other train in that block (a 'clear block') and the train's route is set correctly. In this case, the signal protecting the block is set to green; on the other hand, if a block is not clear, the signal protecting the block is set to red and the signal before is set to yellow (a yellow signal pre-alerts the train driver that the next signal will be red) (Pasquini, et al., 2004).

However, the whole signaling mechanism is designed to be a fail-safe system (signals are set to red in case of problems). When the train fails to stop at a signal set to danger (a red signal), such an event is called signal passed at danger (SPAD). The responsibility for respecting the signal status is assigned to the train driver, as well as a protection system that brakes the train automatically when required.

Indeed, a train may need over a kilometer to stop, since braking should be started at the point at which a yellow signal is reached. The distance a train requires to stop depends on the characteristics of the train, track, braking, and adhesion levels between train and track. That means that drivers must have a high level of attention and vigilance and should get adequate training to understand the trains they operate. The driver behavior can be described as follows: if the signal sets to green, the driver does not have to take any action; if it sets to yellow, the driver must reduce speed to be ready to stop at the next signal; if it sets to red, the driver must stop.

The scenario that will be studied in our case is as follows: a passenger train operating during the day; conditions of weather and visibility at the time of operation are good; no technical failures in the track itself or in the rolling stock, including the brake system. This means that the scope of analysis is limited to train driver errors and failures in the control and signaling system (ERTMS/ETCS levels 1 or 2) that might lead to SPAD. The operation mode is a full supervision mode; in this mode, the on-board ERTMS/ETCS equipment is responsible for train protection.

Identifying Hazards Using STPA

The STPA method will be applied in our case study to identify the hazards that could lead to SPAD in both ERTMS/ETCS operational levels 1 and 2. In fact, this process of using the STPA method will be similar at both operational levels.

The consequences of an accident when SPAD occurs are also the same. These sequences are:

- Death, injury, and/or property damage resulting from a collision with another train.

- Injury or property damage occurring within the train because of incorrect braking technique (without a collision).

As we know, the process of hazard identification using STPA consists of five steps. These steps are as follows:

Hazardous Events

1. The train enters a new block even though the signal is red, which may cause a collision. [H-1]
2. Incorrect braking technique (sudden braking) when there is a red signal for the next block, which subjects the passengers to sudden high forces that may lead to injuries. [H-2]

Safety Constraints

1. The train should not enter the next block when the signal is set to danger (red signal) [SC-1]
2. The brakes should be used correctly (gently) during train braking when the signal is set to danger (red signal) [SC-2]

Hierarchical Safety Control Structure

This structure includes the components of the ERTMS/ETCS levels 1/2 and the interactions between them, as well as the interactions between the components and driver. This structure is shown in Fig 3(a) for ERTMS/ETCS level 1, and Fig 3(b) for level 2.

It can be observed that there is a difference between the functional requirements of the driver and functional requirements of the controller. The functional requirements of the driver are:

- Detect when the line signal is yellow (the next will be red) and activate the brakes;
- Apply the brakes correctly (gently) and ensuring the train stops before the next block.

While the functional requirements of the controller are:

- Detect when there is a need for braking and the driver does not brake;

Figure 3a. Hierarchical safety control structure of ERTMS/ECS level 1

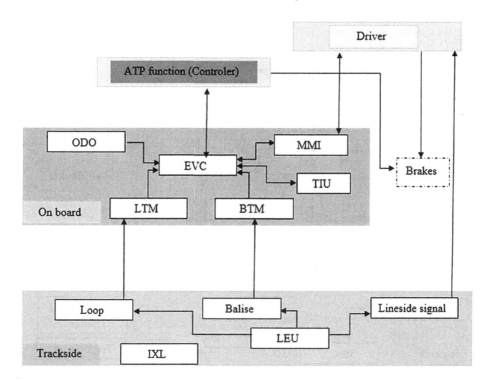

- Respond to a need for braking by applying the braking correctly (gently) and ensuring the train stops before the next block.

Identifying Unsafe Control Actions (UCAs)

The driver is responsible for braking, while the ATP is responsible for protecting the train, meaning that ATP steps in when the braking is not applied by the driver as needed. For the two hazardous events [H-1] and [H-2], the unsafe control actions of both the driver and the ATP will be examined to identify instances in which safety constraints may be violated. UCAs may in fact be caused by various scenarios, including driver errors, component failures, unexpected interactions between components, and missing or delayed feedback from sensors. Table 1 presents UCAs identified for braking actions. The UCAs for both ERTMS/ETCS levels 1 and 2 are identical because the braking action is same, and the difference between these two systems is limited to the communication between onboard and trackside components.

Figure 3b. Hierarchical safety control structure of ERTMS/ECS level 2

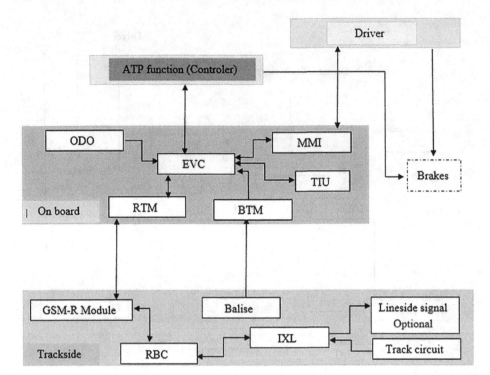

Identifying Causal Scenarios

Here we will look at only some examples (not all) of causal scenarios related to the unsafe control actions mentioned. The UCAs are:

- UCA1: Driver does not brake when the next signal is red.

One of the various scenarios related to UCA1 can be described as follows: UCA1 may occur when the driver fails to brake because he expects that the next signal will be green (does not notice the yellow signal (lineside - level 1, in cab - level 2)).

The first scenario explaining the driver's behavior is that the driver has familiarity with this route and the signal was green at all previous times when he passed it, so the driver thinks the signal will be green this time as well. This belief is formed from his expectation, and he selects the control action depending on his current mental model of the process (knowledge-based decision).

The second scenario for UCA1 is that the driver does not check the status of the signal because he believes the ATP system is active and that there is no failure in the ERTMS/ETCS system. This entails that the ATP system will apply the brakes

Table 1. Unsafe control actions related to braking in ERTMS/ETCS levels 1/2

	Not Providing Causes Hazard	**Providing Causes Hazard**	**Wrong Timing Causes Hazard**	**Stop Too Soon / Applied Too Long**
Brake (Driver)	Driver does not brake when the next signal is red. [H-1]	[Not hazard]	Driver brakes too soon when the next signal is red. [No hazard] Driver waits too long to brake when the next signal is red. [H-2], [H-1].	Driver continues braking for too long when the next signal is red. [No hazard] Driver does not brake for long enough to avoid SPAD when the signal is red. [H-1]
Brake (ATP)	ATP does not brake (when the next signal is red, and driver did not brake). [H-1]	[Not hazard]	ATP brakes too soon (when the next signal is red, and driver did not brake). [Not hazard]. ATP waits too long to brake (when the next signal is red, and driver did not brake). [H-2], [H-1]	ATP continues braking for too long when the next signal is red, and driver did not brake). [not hazard] ATP does not brake for long enough to avoid SPAD (when the signal is red, and driver did not brake). [H-1]

if requires. The driver arrives at this belief because he incorrectly supposes that ERTMS/ETCS is capable of constantly conducting self-diagnosis and will alert him if there is any failure. In reality, however, this is not the case (driver has an incorrect belief about what the system can do).

Many other scenarios can be used to describe the causes of UCA1. A range of factors (such as fatigue, inattention, expectation, and lack of training) could contribute to the occurrence of UCA1.

- UCA2: ATP system does not react appropriately to the driver's behavior.

For level 1, failures can occur if the signal for braking is not sent to ATP (controller) from EVC (because of failure in EVC itself). Also, a UCA2 could result if the command to activate the brakes does not reach the brakes (failure in ATP function (controller) or in the communication channel between the controller and brakes). Moreover, the failure can occur if information about the status of the next block does not reach the EVC at all; this could happen due to a failure in one or more components (track circuit, IXL, LEU, LTM, and BTM) or communication channels (Balise - BTM, Loop - LTM, BTM - EVC, and LTM - EVC).

For level 2, the failure related to EVC and ATP function can occurred in a similar way as for level 1. However, the failure which prevents the status of the next block from reaching the EVC occurs in a different way, either because of the failure in one or more components (track circuit, IXL, RBC, GSM-R module, and RTM) or in one of two communication channels (GSM-R module – RTM and RTM - EVC).

- UCA3: Driver waits too long to brake when the next signal is red.

There are many potential scenarios in which this UCA3 might arise. For example, the driver knows the next signal is red and the ERTMS/ETCS system displays (via MMI) the speed profile and correct point for beginning to brake. However, the driver does not begin to apply the brakes at the correct point because he does not trust the system (incorrect mental model about what the system can do); alternatively, depending on his information about the current state of the system, he may believe incorrectly that characteristics of the train and its brakes allow him to start braking after more time has elapsed (lack of training).

The second scenario is that the driver knows the next signal is red, but he has neither sufficient experience nor training to recognize the information provided by the ERTMS/ETCS system that illustrates the speed profile and determines the point at which braking should commence. Consequently, the driver depends on his limited experience to determine when to apply the brakes (driver has insufficient information about the current state of the system).

The third scenario is that the driver does not recognize the status of the next block because he is affected by fatigue, tiredness, or similar. Consequently, he does not know that the next signal is red at the correct time (wrong belief about the process state).

- UCA4: ATP waits too long to brake when the next signal is red.

For ERTMS/ETCS Level 1, the failure can appear if the data from EVC incorrectly indicates to the ATP (controller) that the train's current speed is less than its actual speed (due to a failure in the EVC itself or feedback about the current speed from the ODO being inaccurate); alternatively, the data indicating the brakes should be applied is sent late to the ATP function (controller), which can occur due to long data processing time in the EVC or a significant delay before sending movement authorities from LTM or BTM to EVC. Another source of failure is a delay in sending a brake activation signal from ATP to the brakes.

For Level 2, this failure can occur in a similar way as for level 1, but the main differences are in the trackside components and communication channel. Thus, the failure can result from a delay in RTM before sending the movement authority to EVC, extended data processing time in IXL and/or RBC before producing the movement authority sent to the GSM-R module, or a delay in communication channels (for example, the channels between the GSM-R module and RTM (GSM protocol) or between the track circuit and IXL (WAN network)).

- UCA5: Driver does not brake for long enough to avoid SPAD when the signal is red.

There are many scenarios associated with UCA5. For example, when the driver sees another train pass his train on the opposite side, he releases the brakes and starts to move the train, passing the signal at red (i.e. does not wait until the signal turns green). The causes of this behavior can be attributed to either the driver's goal to arrive at his destination as soon as possible without making safety his highest priority (lack of safety culture, incorrect policies in the organization), or the driver having an inaccurate mental model about the system, incorrectly believing that the block is free because he saw another train pass him (weak rules and procedures). Alternatively, the driver may believe that the system needs some time to turn the signal green after the block becomes free. However, this is incorrect; the system turns the signal green directly after the block becomes free and switch points are adjusted correctly (wrong belief about system behavior).

- UCA6: ATP does not brake for long enough to avoid SPAD.

For Level 1, the failure can appear either if the data from EVC incorrectly indicates to the ATP function (controller) that the brakes should be released (failure in EVC itself) or a failure in the ATP function leads to release of the brakes. UCA6 can also occur if there is a failure in the track circuit or IXL, leading to incorrect information about the status of the block being sent to EVC (block is free when it is not).

For Level 2, the failure that leads to UCA6 can occur in a similar way to that in level 1, with the addition of a failure in RBC, which could lead to a 'release brakes' command being sent before the block becomes free.

It is worth mentioning that the track circuit is included within the scope of the case study, although it is outside of the scope of the ERTMS system.

Summary of the Results of Hazard Identification

Regarding the different causal scenarios that lead to unsafe control actions, we can classify the causes of hazardous events [H-1], [H-2] for ERTMS/ETCS system operational levels 1 and 2 in a simplified manner, as follows:

- ERTMS/ETCS Level 1.

At this level, the system components and human errors that lead to hazardous events can be divided into three groups: major components, secondary components, and selecting unsafe control actions.

The major components group includes three elements: IXL, Truck circuit, and LEU. A failure in one component of this group (Error1) leads directly to a hazardous event because their functions impact both the driver and ATP. However, if there is no failure in the major components group, the hazardous event will happen only if there are failures in both of the other groups (secondary group and selecting unsafe control actions) (Error2).The components included in the secondary group are Balise, loop, LTM, BTM, ODO, EVC, and ATP function.

While the 'selecting unsafe control actions' group encompasses two elements (i.e. environment and mental model), the environmental element is itself affected by three factors (organization policies, the driver's goals, and time pressure during the driver's decision selection). The mental model element is affected by two subgroups (knowing current data about the system and knowing what the system can do). Many other factors (such as tiredness, fatigue, expectation, familiarity, training, and trust in system) and components (MMI, line signals) impact on these two subgroups.

- ERTMS/ETCS Level 2

The model for Level 1 can be used for Level 2 with minor changes. These changes include the major components and secondary components groups. The major components group in level 2 includes IXL, Truck circuit, RBC, GSM-R module and RTM, while the secondary components group encompasses ODO, EVC, and ATP function. The third group (selecting unsafe control actions) remains the same (only the line signal is neglected). Fig 20 illustrates this model using a Bayesian network.

Since the hazards related to SPAD in railway systems at ERTMS/ETCS system levels 1 and 2 have been identified, we can now suggest a model to represent the hazardous events associated with SPAD, taking into consideration the failure of components and driver errors. By using this model, we can quantify the safety of a railway system in the context of SPAD by using a Bayesian network.

Quantitative Analysis Regarding SPADs with Bayesian Networks

A Bayesian network (BN) is a very effective method to map out all types of dependencies and illustrate the relationships between causes and final outcomes (i.e. hazardous events) in a system using a graphical model. A BN model is built in three steps, as follows:

1. Building a BN diagram

This diagram consists of a directed acyclic graph showing the dependencies among the different factors by using arrows (links). In the presenting model, a driver's selection of a control action is impacted by many factors, such as tiredness, training, trust in system, etc. The action taken can be safe or unsafe; if unsafe, it can be corrected by the ATP system.

2. Determining the states of root nodes and their probabilities

In our case, for simplicity, we will assume that all nodes have only two possible states, i.e. 1 and 0. Thus, node $(X) = 1$ if the event represented by this node happens and node $(X) = 0$ if the event does not happen. The probabilities of the events occurring over a period of 10^6 hours, which are represented by root nodes, are provided in Table 2.

3. Determining the conditional probability tables of other nodes

As we know, conditional probability tables are used to determine the probability of the states of each node. In our case, the values of these tables are arbitrary (experts or data analysis were not relied upon to determine the values of these tables). Since the number of parents of nodes determines the size of the conditional probability table for this node, the table size in our case will be no more than 4 or 8 rows, as the maximum number of parents is 3.

4. Calculating the probability of hazardous events

Calculation results will depend on the values of the states of root nodes and the values in the conditional probability tables. In this case, Excel will be used to conduct this calculation.

Fig 5 shows the model and related conditional probabilities in the quantitative safety analysis of ERTMS/ETCS level 1 related to SPAD. The probability of a correct mental model is 0.653546164, while the probability of an incorrect mental model is 0.346453836. The probability of a good environment is 0.74752672, while the probability of an environment that is not good is 0.25247328. The probability of selecting a safe control action is 0.719703131, while the probability of selecting an unsafe control action is 0.280296869.

Figure 4a. BNs of hazardous events related to SPAD in ERTMS/ETCS level 1

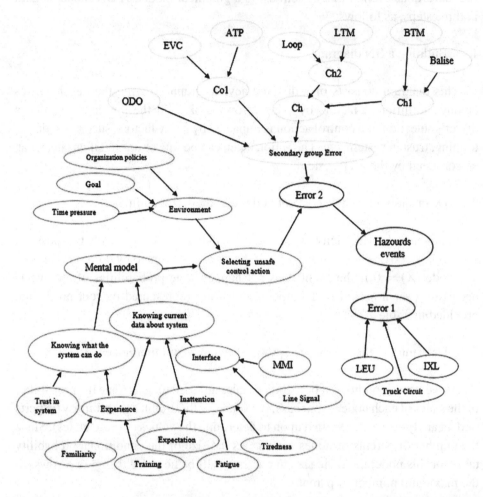

Use the similar approach, we can obtain that the probability of secondary group error is 0.401058786, the probability of Error 2 is 0.112415522, the probability of Error 1 is 0.058808, and the probability of a hazardous event is 0.16461259 More details about the analysis can be found in the master thesis by Rekabi (2018).

The quantitative safety analysis for ERTMS/ETCS level 2 is similar to that for level 1. Because level 2 involves continuous supervision of train movement with continuous communication, which is provided by GSM-R, this leads to a reduction in human errors, as the data will be available to the driver for a longer time, thus reducing the inattention and time pressure and giving the driver more time to make a correct decision. In our ERTMS/ETCS level 2 model, lineside signals are omitted. Fig. 6 shows the mental model and conditional probability tables of ERTMS/ETCS Level 2.

Figure 4b. BNs of hazardous events related to SPAD in ERTMS/ETCS level 2

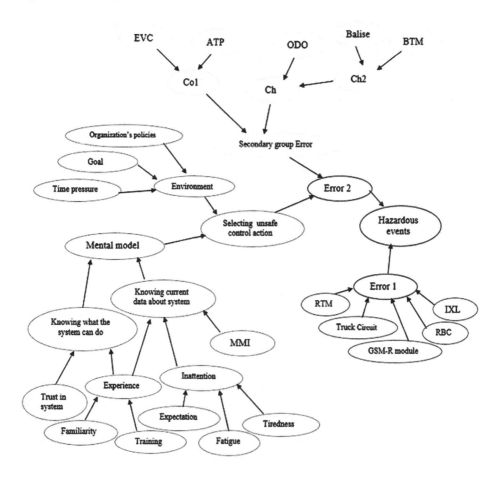

The probability of a correct mental model is 0.866879, while the probability of an incorrect mental model is 0.133121. The probability of a good environment is 0.806337, while the probability of an environment that is not good is 0.193663. The probability of selecting a safe control action is 0.879917, while the probability of selecting an unsafe control action is 0.120083. The probability of secondary group error is 0.052032, the probability of Error 2 is 0.006248, the probability of Error 1 is 0.096079, and the probability of a hazardous event is 0.101727.

The comparison between the results of the quantitative safety analysis of ERTMS/ETCS levels 1 and 2 related to SPAD are shown in table 3.

We can observe that ERTMS level 2 is safer than ERTMS level 1. The improvement in safety comes from keeping the train driver continuously informed about the states of signals ahead through RBC and GSM-R systems. The driver has enough time to

Table 2. Probability of each state of root nodes in ERTMS/ETCS levels 1 and 2

The Variable (Node)	Not Happened Pr (X)	Happened Pr (X)	The Event (Node)	Not Happened Pr (X)	Happened Pr (X)
Tiredness	0.90	0.10	EVC (Failure)	0.98	0.02
Fatigue	0.85	0.15	ATP (Failure)	0.98	0.02
Expectation (bad impact)	0.90	0.10	ODO (Failure)	0.95	0.05
Training	0.15	0.85	Balise (Failure)	0.98	0.02
Familiarity	0.30	0.70	BTM (Failure)	0.98	0.02
Trust in system	0.15	0.85	RTM (Failure)	0.98	0.02
Goal (bad impact)	0.88	0.12	RBC (Failure)	0.98	0.02
LTM, LEU (Failure)	0.98	0.02	Loop (Failure)	0.98	0.02
IXL (Failure)	0.98	0.02	Truck circuit (Failure)	0.98	0.02
Time pressure	0.97	0.03	GSM-R Module (Failure)	0.98	0.02
MMI (Failure)	0.95	0.05			
Policies of organization (Bad impact)	0.86	0.14	Line signal (Failure)	0.96	0.04

make the correct decision, thus reducing the probability of errors. At level 1, on the other hand, the driver is informed about the states of signals ahead only when he looks at the lineside signals, which can be affected by many factors related to the state of the driver and the weather conditions. Also, we can observe the probability of driver errors (selecting unsafe control actions) in ERTMS level 2 is lower than for level 1 for the same reason.

On the other hand, the number of critical components in ERTMS level 2 is higher than in level 1. Therefore, the probability of errors being caused by these components (error 1) is higher in level 2 than level 1. The reliability of these components is thus a critical issue and should always be as high as possible.

Sensitivity Analysis

The aim of sensitive analysis in our model is to determine how the output value (which presents the probability of hazardous event SPAD) can be changed when we change probability of states of one variable (factor), so that we can make the probability of ideal value is (1) for one variable and keep the probability of the rest of the variables. We then repeat this process with different variables (factors) and copy the value of the probability of hazardous events for each variable. Completing

Figure 5. Model of mental model using Bayesian network and related conditional probability tables – ERTMS/ETCS level 1

Tiredness	Fatigue	Expectation (bad impact)	Inattention	
			Pr (X)	Pr (X)
1	1	1	0.01	0.99
1	1	0	0.10	0.90
1	0	1	0.30	0.70
1	0	0	0.35	0.65
0	1	1	0.40	0.60
0	1	0	0.50	0.50
0	0	1	0.60	0.40
0	0	0	0.80	0.20

Training	Familiarity	Experience	
		Pr (X)	Pr (X)
1	1	0.05	0.95
1	0	0.25	0.75
0	1	0.40	0.60
0	0	0.85	0.15

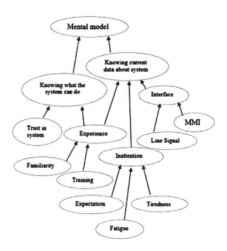

Experience	Trust in system	Knowing what the system can do	
		Pr (X)	Pr (X)
1	1	0.20	0.80
1	0	0.35	0.65
0	1	0.40	0.60
0	0	0.95	0.05

MMI (failure)	Line signal (failure)	Interface (failure)	
		Pr (X)	Pr (X)
1	1	0.00	1.00
1	0	0.20	0.80
0	1	0.30	0.70
0	0	0.95	0.05

Inattention	Interface (Failure)	Experience	Knowing current data about system	
			Pr (X)	Pr (X)
1	1	1	0.99	0.01
1	1	0	1.00	0.00
1	0	1	0.93	0.07
1	0	0	0.98	0.02
0	1	1	0.95	0.05
0	1	0	0.99	0.01
0	0	1	0.15	0.85
0	0	0	0.60	0.40

Knowing current data about system	Knowing what the system can do	Mental model	
		Pr (X)	Pr (X)
1	1	0.00	1.00
1	0	0.35	0.65
0	1	0.45	0.55
0	0	0.99	0.01

this process will yield Table 4, which gives us the values of hazardous events for ERTMS/ETCS level 1 and level 2 when we use ideal values with one variable.

It can be found that

- The best improvement in system safety is obtained when we cause the driver's objective (goal) to align with the 'safety first' concept.
- EVC is the most critical physical component in the system, as increasing its reliability has the greatest impact on improving the safety of the system as a whole.

Figure 6. Mental model using Bayesian network and related conditional probability tables - ERTMS level 2

Tiredness	Fatigue	Expectation (bad impact)	Inattention	
			Pr (X)	Pr (X)
1	1	1	0.15	0.85
1	1	0	0.30	0.70
1	0	1	0.45	0.55
1	0	0	0.50	0.50
0	1	1	0.55	0.45
0	1	0	0.70	0.30
0	0	1	0.80	0.20
0	0	0	0.90	0.10

Training	Familiarity	Experience	
		Pr (X)	Pr (X)
1	1	0.05	0.95
1	0	0.25	0.75
0	1	0.40	0.60
0	0	0.85	0.15

Inattention	MMI (Failure)	Experience	Knowing current data about system	
			Pr (X)	Pr (X)
1	1	1	0.95	0.05
1	1	0	1.00	0.00
1	0	1	0.85	0.15
1	0	0	0.92	0.08
0	1	1	0.80	0.20
0	1	0	0.85	0.15
0	0	1	0.00	1.00
0	0	0	0.70	0.30

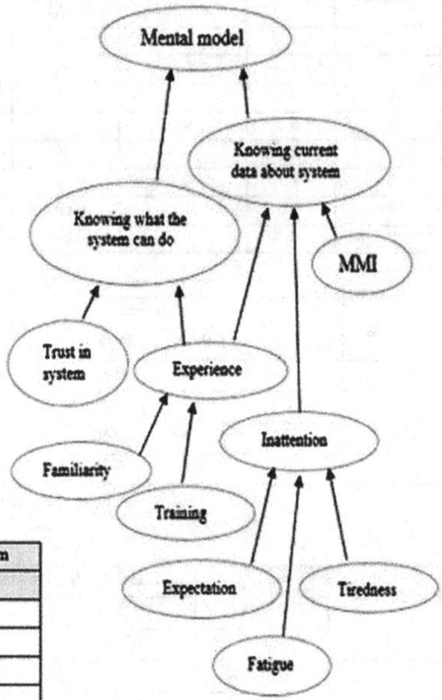

Experience	Trust in system	Knowing what the system can do	
		Pr (X)	Pr (X)
1	1	0.10	0.90
1	0	0.25	0.75
0	1	0.15	0.85
0	0	0.60	0.40

Knowing current data about system	Knowing what the system can do	Mental model	
		Pr (X)	Pr (X)
1	1	0.00	1.00
1	0	0.25	0.75
0	1	0.30	0.70
0	0	0.90	0.10

Table 3. Comparison of the safety between ERTMS/ETCS level 1 & 2 related to SPAD

ERTMS/ETSC	Wrong Mental Model	Selecting Unsafe Control Action	Error 2	Error 1	Hazardous Event
Level 1	0.346454	0.280297	0.112416	0.058808	0.164613
Level 2	0.133121	0.120083	0.006248	0.096079	0.101727

A wide range of factors can potentially influence the train driver's mental model, along with his ability to select the correct action. It should be noted that only some of these factors were incorporated in our suggested model in order to keep the model as simple as possible. Moreover, the results of our calculation are approximate and not accurate, as we were unable to obtain accurate data from experts and accident investigations; consequently, we used estimated values.

CONCLUSION AND PERSPECTIVES

Most train drivers' tasks in the ERTMS system are mental tasks and are thus exposed to errors emerging from flaws in the mental process model or in selection control action. Therefore, analyzing safety in the ERTMS system in relation to train drivers depends heavily on new techniques of risk analysis, such as the STPA method. This method can be used effectively to identify all types of hazards associated with train drivers' tasks. Moreover, given its benefits, BN is considered a suitable method for modelling and quantifying the different errors that can lead to hazardous events.

A quantitative safety analysis related to SPAD in ERTMS system levels 1 and 2 (including human errors) was performed as a case study to assess and compare safety between those two levels. The results of this analysis show that ERTMS level 2 is

Table 4. Probabilities of hazardous events related to SPAD when the probabilities of some variables are 1

The Variable (Node)	Probability Happened Pr(X)	ERTMS/ETCS Probabilities of Hazardous Event	
		Level 1	Level 2
Tiredness	0	0.162279	0.101596
Fatigue	0	0.162279	0.101625
Training	1	0.161755	0.101534
Familiarity	1	0.161420	0.101511
Trust in system	1	0.162103	0.101586
Goal (Bad impact)	0	0.151651	0.100659
Organization polices	1	0.156933	0.101190
MMI (Failure)	0	0.162978	0.101602
Blaise (Failure)	0	0.161748	0.101446
RBC (Failure)	0	--	0.101842
EVC (Failure)	0	0.161388	0.099627
The original value		0.164613	0.101727

safer and less prone to driver errors than ERTMS level 1; however, it also contains more critical elements (such as GSM-R system and RBC) that have a significant impact on the continuity of ERTMS functioning, such that any failure in one of these components will stop the whole ERTMS system.

Conditional probability tables (CPTs) are very important and have a notable impact on the outcome of the Bayesian network. Various studies can be dedicated to identifying values of conditional probability tables for all factors in the safety analysis model; these values can then be considered as standards to be used in different countries and organizations to mitigate the variability in the safety analysis results.

REFERENCES

Barger, P., Schön, W., & Bouali, M. (2009). *A study of railway ERTMS safety with Colored Petri Nets.* The European Safety and Reliability Conference 2009, Prague, Czech Republic.

Baysari, M. T., Caponecchia, C., McIntosh, A. S., & Wilson, J. R. (2009). Classification of errors contributing to rail incidents and accidents: A comparison of two human error identification techniques. *Safety Science*, *47*(7), 948–957. doi:10.1016/j.ssci.2008.09.012

Cacciabue, P. C. (2005). Human error risk management methodology for safety audit of a large railway organisation. *Applied Ergonomics*, *36*(6), 709–718. doi:10.1016/j.apergo.2005.04.005 PMID:16122693

Callet, S., Fassi, S., Fedeler, H., Ledoux, D., & Navarro, T. (2014). *The use of a "model-based design" approach on an ERTMS Level 2 ground system. In Formal Methods Applied to Industrial Complex Systems* (pp. 165–190). Hoboken, NJ: John Wiley & Sons.

Castillo, E., Calviño, A., Grande, Z., Sánchez-Cambronero, S., Rivas, I. G. A., & Menéndez, J. M. (2016). A Markovian–Bayesian network for risk analysis of high speed and conventional railway lines integrating human errors. *Computer-Aided Civil and Infrastructure Engineering*, *31*(3), 193–218. doi:10.1111/mice.12153

Castillo, E., Grande, Z., Mora, E., Lo, H. K., & Xu, X. D. (2017). Complexity reduction and sensitivity analysis in road probabilistic safety assessment Bayesian network models. *Computer-Aided Civil and Infrastructure Engineering*, *32*(7), 546–561. doi:10.1111/mice.12273

Dorrian, J., Roach, G. D., Fletcher, A., & Dawson, D. (2007). Simulated train driving: Fatigue, self-awareness and cognitive disengagement. *Applied Ergonomics, 38*(2), 155–166. doi:10.1016/j.apergo.2006.03.006 PMID:16854365

Felipe, A., Sallak, M., Schön, W., & Belmonte, F. (2013). Application of evidential networks in quantitative analysis of railway accidents. *Proceedings of the Institution of Mechanical Engineers, Part O: Journal of Risk and Reliability, 227*(4), 368–384.

Flammini, F., Marrone, S., Iacono, M., Mazzocca, N., & Vittorini, V. (2014). A Multiformalism modular approach to ERTMS/ETCS failure modeling. *International Journal of Reliability Quality and Safety Engineering, 21*(1), 1450001. doi:10.1142/S0218539314500016

Flammini, F., Marrone, S., & Mazzocca, N. (2006). *Modelling system reliability aspects of ERTMS/ETCS by fault trees and Bayesian networks*. The European Safety and Reliability Conference 2006, Estoril, Portugal.

France, M. E. (2017) *Engineering for Humans: A New Extension to STPA, in Aeronautics and Astronautics* (Master thesis). Massachusetts Institute of Technology.

Ghazel, M. (2014). Formalizing a subset of ERTMS/ETCS specifications for verification purposes. *Transportation Research Part C: Engineering Technologies, 42*, 60–75. doi:10.1016/j.trc.2014.02.002

Hamilton, W. I., & Clarke, T. (2005). Driver performance modelling and its practical application to railway safety. *Applied Ergonomics, 36*(6), 661–670. doi:10.1016/j.apergo.2005.07.005 PMID:16182232

Kyriakidis, M., Majumdar, A., & Ochieng, W. (2018). The human performance railway operational index-a novel approach to assess human performance for railway operations. *Reliability Engineering & System Safety, 170*, 226–243. doi:10.1016/j.ress.2017.10.012

Kyriakidis, M., Pak, K. T., & Majumdar, A. (2015). Railway accidents caused by human error: Historic analysis of UK railways, 1945 to 2012. *Transportation Research Record: Journal of the Transportation Research Board, 2476*(1), 126–136. doi:10.3141/2476-17

Lee, S.-H., & Song, J. (2016). Bayesian-network-based system identification of spatial distribution of structural parameters. *Engineering Structures, 127*, 260–277. doi:10.1016/j.engstruct.2016.08.029

Leveson, N. (2012). *Engineering a Safer World: Systems Thinking Applied to Safety*. Cambridge, MA: MIT Press. doi:10.7551/mitpress/8179.001.0001

Marseguerra, M., Zio, E., & Librizzi, M. (2006). Quantitative developments in the cognitive reliability and error analysis method (CREAM) for the assessment of human performance. *Annals of Nuclear Energy, 33*(10), 894–910. doi:10.1016/j.anucene.2006.05.003

McLeod, R. W., Walker, G. H., & Moray, N. (2005). Analysing and modelling train driver performance. *Applied Ergonomics, 36*(6), 671–680. doi:10.1016/j.apergo.2005.05.006 PMID:16095554

Mu, L., Xiao, B. P., Xue, W. K., & Yuan, Z. (2015) The prediction of human error probability based on Bayesian networks in the process of task. *2015 IEEE International Conference on Industrial Engineering and Engineering Management (IEEM)*. 10.1109/IEEM.2015.7385625

Pasquini, A., Rizzo, A., & Save, L. (2004). A methodology for the analysis of SPAD. *Safety Science, 42*(5), 437–455. doi:10.1016/j.ssci.2003.09.010

Plioutsias, A., & Karanikas, N. (2015). Using STPA in the evaluation of fighter pilots training programs. *Procedia Engineering, 128*, 25–34. doi:10.1016/j.proeng.2015.11.501

Rausand, M. (2011). *Risk Assessment: Theory, Methods, and Applications*. John Wiley & Sons. doi:10.1002/9781118281116

Rekabi, M. M. (2018). *Bayesian Safety Analysis of Railway Systems with Driver Errors* (Master thesis). Norwegian University of Science and Technology.

Wilson, J. R., Farrington-Darby, T., Cox, G., Bye, R., & Hockey, G. R. J. (2007). The railway as a socio-technical system: Human factors at the heart of successful rail engineering. *Proceedings of the Institution of Mechanical Engineers. Part F, Journal of Rail and Rapid Transit, 221*(1), 101–115. doi:10.1243/09544097JRRT78

Chapter 7
Performability Modeling of Distributed Systems and Its Formal Methods Representation

Razib Hayat Khan
American International University-Bangladesh, Bangladesh, Dhaka

ABSTRACT

A distributed system is a complex system. Developing complex systems is a demanding task when attempting to achieve functional and non-functional properties such as synchronization, communication, fault tolerance. These properties impose immense complexities on the design, development, and implementation of the system that incur massive effort and cost. Therefore, it is vital to ensure that the system must satisfy the functional and non-functional properties. Once a distributed system is developed, it is very difficult and demanding to conduct any modification in its architecture. As a result, the quantitative analysis of a complex distributed system at the early stage of the development process is always an essential and intricate endeavor. To meet the above challenge, this chapter introduces an extensive framework for performability evaluation of a distributed system. The goal of the performability modeling framework is to consider the behavioral change of the system components due to failures. This reveals how such behavioral changes affect the system performance.

DOI: 10.4018/978-1-7998-1718-5.ch007

INTRODUCTION

Conducting performance evaluation of a distributed system separately from the dependability evaluation fails to assess the anticipated system performance in the presence of system components failure and recovery. System dynamics is affected by any state changes of the system components due to failure and recovery. This chapter introduces the concept of performability that considers the behavioral change of the system components due to failures and also reveals how this behavioral change affects the system performance. But to design a composite model for a distributed system, perfect modeling of the overall system behavior is essential and sometimes very unwieldy. A distributed system behavior is normally realized by the several objects that are physically disseminated. The overall system behavior is maintained by the partial behavior of the distributed objects of the system (Khan & Heegaard, 2011). So, it is essential to model the distributed objects behavior perfectly for appropriate demonstration of the system dynamics and to conduct the performability evaluation (Khan & Heegaard, 2011). Hence, we adopt Unified Modeling Language (UML) collaboration, state machine, deployment, and activity-oriented approach as UML is the most commonly used specification language which models both the system requirements and qualitative behavior through an assortment of notations (OMG, 2017; Khan & Heegaard, 2011). Our way of utilizing UML collaboration and activity diagram to capture the system dynamics, provides the opportunity to reuse the software components. The specifications of collaboration are given as coherent, self-contained building blocks (Khan & Heegaard, 2011). Reusability of the software component is achieved by designing the collaborative building block, which is used as main specification unit in this work.

In order to guarantee the precise understanding and correctness of the model, the approach requires formal reasoning on the semantics of the language used and to maintain the consistency of the models' specification. Temporal logic is a suitable option in this regard. In particular, the properties of super position supported by compositional temporal logic of action (cTLA) make it possible to describe systems from different viewpoints and by individual processes that are superimposed (Herrmann & Krumm, 2000). In this work, we focus on the cTLA that allows us formalizing the collaborative service specifications given by UML activities, deployment diagram and state machine (STM) model. By expressing collaborations as cTLA processes, we can ensure that a composed service maintains the properties of the individual collaborations it is comprised of. The semantic definition of collaboration, activity, deployment, and STM model in the form of temporal logic is implemented as a transformation tool which produces TLA$^+$ modules. These modules may then be used as input for the temporal model checker (TLC) for syntactic analysis (Slåtten, 2007).

Furthermore, UML models are annotated according to the UML profile for Modeling and Analysis of Real- Time and Embedded Systems (MARTE) (OMG 2019) and UML profile for Modeling Quality of Service and Fault Tolerance Characteristics (OMG, 2008) to include quantitative system parameters necessary for performability evaluation. Considering the annotated UML models, UML specification styles are applied to generate SRN models automatically following the model transformation rules. The execution of performance SRN is dependent on the execution of dependability SRN. A transition in the dependability SRN may induce a state change in the performance SRN. The stepwise process to generate the performability SRN model is covered in this chapter.

Over decades several performability modeling techniques have been considered such as Markov models, Stochastic Petri Net (SPN), and Stochastic Reward Net (SRN) (Trivedi, 2011). Among all of these, we will focus on the SRN as performability model generated by our framework due to its prominent and interesting properties such as priorities assignment in transitions, presence of guard functions for enabling transitions that can use entire state of the net rather than a particular state, marking dependent arc multiplicity that can change the structure of the net, marking dependent firing rates, and reward rates defined at the net level.

The modeling process is supported by a set of tools, including Arctis and SHARPE with the incremental model checking facility (Khan, 2014). Arctis is used for specifying the system functional behavior. The evaluation of the performability models generated by the framework is achieved using SHARPE.

The objective of this chapter is to describe our novel approach that allows the evaluation of the performance and dependability related behavior of a distributed systems in a combined and automated way. The chapter is organized as follows: Next section defines the literature study, subsequent sections introduce our performability modeling framework, depict UML model description, explain service components deployment mapping issue, describe formalization of UML models, delineate model transformation rules, introduce the model synchronization mechanism, describe the hierarchical method for mean time to failure (MTTF) calculation, illustrate the case study, and finally, it delineates the concluding remarks with future directions.

LITERATURE REVIEW

Several approaches have been pursued to accomplish a performability analysis model from a system design specification. Sato et al. develop a set of Markov models, for computing the performance and the reliability of Web services and detecting bottlenecks (Sato & Trivedi, 2007). Another initiative focuses on model-based analysis of performability of mobile software systems by proposing a general methodology that starts from design artifacts expressed in a UML-based notation. Inferred performability models are formed based on the Stochastic Activity Networks (SAN) notation (Bracchi, Cukic, & Cortellesa, 2014). Subsequent effort proposes a methodology for the modeling, verification, and performance evaluation of communication components of a distributed application building software which translates UML 2.0 specifications into executable simulation models (Wet & Kritzinger, (2011). Gonczy et al. mentioned a method for high-level UML models of service configurations captured by a UML profile dedicated to service design; performability models are derived by automated model transformations for the PEPA toolkit in order to assess the cost of fault tolerance techniques in terms of performance (Deri & Varro, 2008). However, most of the existing approaches do not consider the fact of how to conduct the system modeling to delineate system functional behavior while generating the performability model using reusable software components. The framework introduced in this work is superior to the existing approaches that have been realized by UML specification style as reusable building block to characterize system dynamics. The purpose of the reusable building block is twofold: to express the local behavior of several components and to capture the interaction between them. This provides the excellent opportunity to reuse the building blocks, as the interaction among the several components can be encapsulated within one self-contained building block (Khan & Heegaard, 2011). This reusability provides the means to design a new system's behavior rapidly utilizing the existing building blocks according to the specification. This helps to start the development process not from scratch which in turn facilitates the swelling of productivity and quality in accordance with the reduction in time and cost (Khan, Fumio, Heegaard, & Trivedi, 2012b). Moreover, the ensuing deployment mapping given by our framework has greater impact to satisfy quality of service (QoS) requirements provided by the system. The target in this work is to deal with vector of QoS instead of confining them in one dimension. Our provided deployment logic is definitely capable of handling any properties of the service as long as a cost function for the specific property can be produced. The defined cost function is able to react in accordance with the changing size of search space of available hosts presented in the execution environment to assure an efficient deployment mapping (Khan & Heegaard, 2011). In addition, the separation of performance and dependability model and the introduction of model synchronization to synchronize the two models' activities using guard functions relinquishes the complex and unwieldy affect in performability modeling and evaluation of large and multifaceted systems (Khan, Fumio, Heegaard, & Trivedi, 2012a).

OVERVIEWS OF THE FRAMEWORK

Considering the above-mentioned factors, the general structure of the performability modeling framework for a distributed system is illustrated in Figure 1 where the service is evaluated by an analytic approach SRN. The same framework can be applied with simulations, but in this chapter, the focus is on analytic models. The rounded rectangle in the Figure 1 represents operational steps, whereas the square boxes represent input/output data. The inputs for the automated model transformation process are as follows:

- A system or service functional behavior specification
- Information regarding system execution environment
- Non-functional parameters for quantitative analysis.

The model representations of the functional behavior, physical platform, and non- functional properties are combined to form annotated models using the UML profiles mentioned above. Then, a given deployment of the service components onto the currently available physical resource (assumed static throughout the evaluation) is added, using annotated models. Finally, this deployment specification is automatically translated into an SRN model, where the performability of the services can be evaluated. If necessary, the evaluation results can be fed into the system design model to identify the performance anti-patterns that might cause problems in system performance.

For ease of understanding the complexity behind the modeling of performability attributes, our modeling framework works in two different views such as performance modeling view and dependability modeling view which is shown in Figure 2. The framework achieves its objective by maintaining harmonization between performance and dependability modeling view with the support of model synchronization.

The service specification models of our modeling framework, such as the UML collaboration and activity diagram, are generated using the Arctis tool (Khan, 2014). Arctis focuses on the abstract, reusable service specifications that are composed of UML collaborations and activities. It uses collaborative building blocks as reusable specification units to provide the structural and behavioral aspects of the service components. In order to support the construction of building blocks, Arctis offers special actions and wizards. Arctis provides an editor to specify services, which allows the user to create collaborations from scratch or combine existing ones taken from a library to create composite collaborations.

SHARPE is a tool that accepts specifications of mathematical models and requests for model analysis (Trivedi & Sahner, 2002). It is a tool to specify and analyze performance, reliability, and performability models. It is a toolkit that provides a specification language and solution methods for most of the commonly used model types. Non-functional requirements of the distributed system can also be evaluated using SHARPE, such as response time, throughput, job success probability, etc.

When using the framework, several feedback loops might exist, depending on the objective of the case study. The feedbacks are as follows:

- System evaluation results can be utilized to identify any discrepancy in the model that demonstrates system functional behavior (feedback from "Evaluation result" to "Service functional behavior").
- The deployment mapping might reveal that there is no feasible solution, forcing the alteration of the physical infrastructure (feedback from "Deployment mapping" to "System physical platform").
- Different deployment strategies can be attempted on the same physical platform with the same service components, and then, the automated translation to analytic model and corresponding assessment is conducted for each deployment (feedback from "Evaluation result" to "Deployment mapping").
- Performability parameters sensitivity can also be checked for different resource constraints in the physical infrastructure (feedback from "Evaluation result" to "System physical platform").

UML BASED SYSTEM DESCRIPTION

UML based system description of our framework is described in the below subsections:

Figure 1.

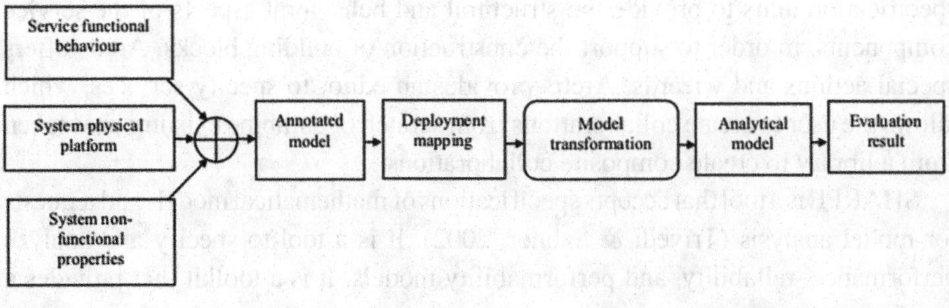

208

Figure 2. Performability Modeling Framework

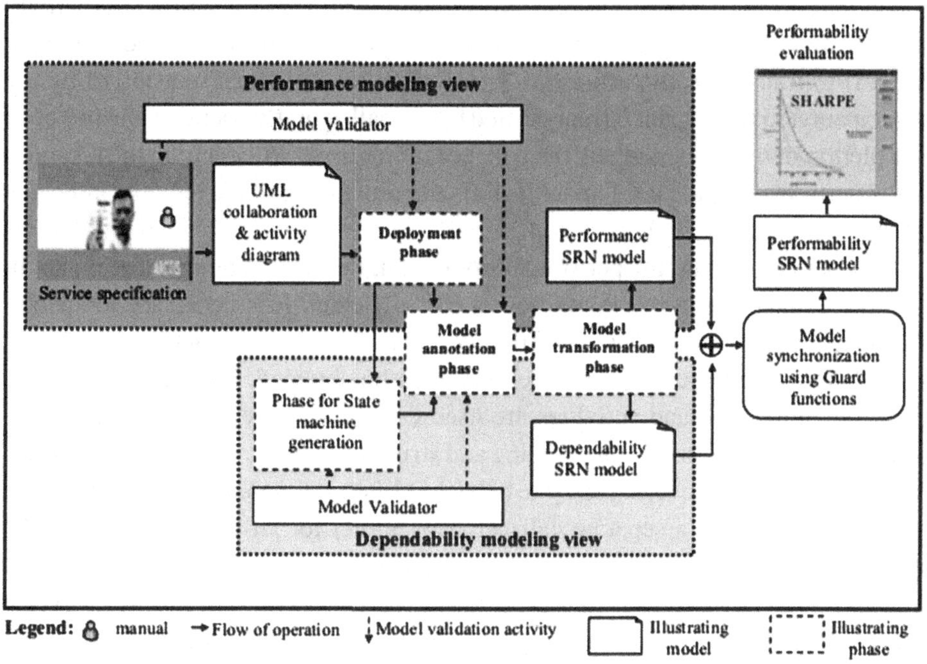

Construction of Collaborative Building Blocks

The performability modeling framework utilizes collaboration as main entity. Collaboration is an illustration of the relationship and interaction among software objects in the UML. Objects are shown as rectangles with naming label inside. The relationships between the objects are shown in an oval connecting the rectangles (OMG, 2017). The specifications for collaborations are given as coherent, self-contained reusable building blocks. The structure of the building block is described by UML 2.5 collaboration. The building block declares the participants (as collaboration roles) and connection between them. The internal behavior of building block is described by the UML activity. It is declared as the classifier behavior of the collaboration and has one activity partition for each collaboration role in the structural description. For each collaboration, the activity declares a corresponding call behavior action referring to the activities of the employed building blocks. For example, the general structure of the building block t is given in Figure 3 where it only declares the participants A and B as collaboration roles and the connection between them is defined as collaboration t_x $(x=1...n_{AB}$ (number of collaborations between collaboration roles A & B)). The internal behavior of the same building block is shown in Figure 4(b). The activity $transfer_{ij}$ (where $ij = AB$) describes the behavior of the corresponding collaboration. It has one activity partition for each collaboration role: *A* and *B*. Activities base their semantics on token flow. The activity starts by forwarding a token when there is a response (indicated by the streaming pin *res*) to transfer from participant *A* to *B*. The token is then transferred by the participant *A* to participant *B* (represented by the call operation action *forward*) after completion of the processing by the collaboration role *A*. After getting the response of the participant *A*, the participant *B* starts the processing of the request (indicated by the streaming pin *req*).

In order to generate the performability model, the structural information about how the collaborations are composed is not sufficient. It is necessary to specify the detailed behavior of how the different events of collaborations are composed so that the desired overall system behavior can be obtained. For the composition, UML collaborations and activities are used complementary to each other. UML collaborations focus on the role binding and structural aspect, while UML activities complement this by covering also the behavioral aspect for composition. Therefore, the activity contains a separate call behavior action for all collaborations of the system. Collaboration is represented by connecting their input and output pins. Arbitrary logic between pins may be used to synchronize the building block events and transfer data between them. By connecting the individual input and output pins of the call behavior actions, the events occurring in different collaborations can be coupled with each other. Semantics of the different kinds of pins are given in more details in (Khan & Heegaard, 2011). For example, the detail behavior and composition of the collaboration is given in following Figure 4(a). The initial node (•) indicates the starting of the activity. The activity is started from the participant *A*. After being activated, each participant starts its processing of request which is mentioned by call operation action Pr_i *(Processing₀, where i = A, B, & C)*. Completion of the processing by the participants are mentioned by the call operation action Prd_i *(Processing_done₀, where i = A, B, & C)*. After completion of the processing, the response is delivered to the corresponding participant. When the processing of the task by the participant *A* completes, the response (indicated by streaming pin *res)* is transferred to the participant *B* mentioned by collaboration *t: transfer$_{ij}$ (where ij = AB)* and participant *B* starts the processing of the request (indicated by streaming pin *req)*. After completion of the processing, participant *B* transfers the response to the participant *C* mentioned by collaboration *t: transfer$_{ij}$ (where ij = BC)*. Participant *C* starts the processing after receiving the response from *B* and activity is terminated after completion of the processing which is illustrated by the terminating node (•).

UML Deployment Diagram

The static view of the system physical platform is illustrated using a UML deployment diagram. A UML deployment diagram is used in our modeling framework to define the execution architecture of the system by identifying the system physical components, the connection between physical components, and the assignment of software artifacts to those identified physical components (OMG, 2017). Service delivered by the system is defined by the joint behavior of the system components, which are physically distributed. This, in turn, aids in exposing of the direct mapping between the software components to the system physical components to exhibit the probable deployment of the service.

Modeling Failure and Repair Behavior of Software and Hardware Component Using UML STM

State transitions of a system element are described using UML State Machine (STM) diagram. In an STM, a state is depicted as a rectangle and a transition from one state to another is represented by an arrow (OMG, 2017). In this work, STM is used to describe the failure and recovery behavior of software and hardware components. The STM of hardware component is shown in Figure 5(a). The initial node (●)

Figure 3. Collaborative Building Block

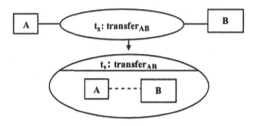

Figure 4. Collaborative Building Block Realized by Activity Diagram

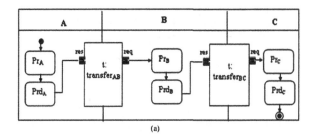

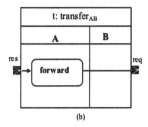

indicates the starting of the operation of hardware component. Then the component enters running state. Running is the only available state here. If the active component fails during the operation and the hot standby component is available, the standby component will take charge and the component operation will be continued. When any failure (whether active component or standby component) incurs, the recovery operation will be performed.

DEPLOYMENT MAPPING

The allocation of software components to the available physical resources of the distributed system is defined as deployment mapping, which has a considerable impact on the desired QoS provided by the system. We model the system as a collection of N interconnected physical nodes. Our objective is to find a deployment mapping for this execution environment for a set of service components that comprises the service. Deployment mapping M is defined as $M=(C\rightarrow N)$, between a number of service components instances $\{c_1, c_2, ...\}$ ε C (captured by collaboration diagram), onto physical nodes $\{n_1, n_2, ...\}$ ε N (captured by UML deployment diagram). In this setting, the service components communicate with each other via a set of collaborations $\{k_1, k_2, ...\}$ ε K. Hence, a collaboration k_j may exist between two components c_a and c_b.

FORMAL METHOD REPRESENTATION OF UML MODEL

The UML specification style introduced in this chapter has been utilized to define the exact and complete behavior of the service specification that focuses on the functionalities they offer and will be used as an input model for the model

Figure 5. UML STM Diagrams of Software and Hardware Components

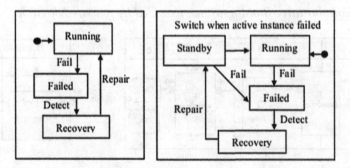

transformation process. Though UML provides comprehensive architectural modeling capabilities, it lacks the ability to formally present the modeling specifications and does not convey formal semantics or syntax. As a result, we delineate the precise semantics of UML collaborations, activities, and deployment by formalizing the concept in the temporal logic cTLA style, which is defined as cTLA/c and the semantics of state machine diagram realized by the cTLA formula Khan, Machida, Heegaard, & Trivedi, 2012b. The motivation behind expressing the semantics using cTLA is to describe various forms of structures and actions through an assortment of operators and techniques, which correspond superbly with UML collaborations, activities, deployment, and the state machine diagram.

cTLA/c for Collaborative Service Specification

cTLA/c is a formal basis for defining the collaborative service specification using UML collaboration and activity (Khan & Heegaard, 2014). The concept of UML collaboration introduced here is rather structural and describes a structure of collaborating elements. In order to illustrate the structural concept of the collaboration, collaborations are mapped into a cTLA/c process, where the process is realized between the collaboration roles internal to the collaborations. In (Khan & Heegaard, 2014), the detailed process for formalizing the UML collaboration specification is introduced, where the focus iss not only to specify the behavior internally to the collaboration but also to define the mechanism to couple the collaborations with others during the composition if necessary.

UML activities have been utilized to express the behavior of collaborations. A UML collaboration is complemented by an activity, which uses one separate activity partition for each collaboration role. In terms of the cTLA/c, an activity partition corresponds to a collaboration role. The semantics of UML activities are based on the Petri nets OMG, 2017. Thus, an activity essentially describes a state transition system, with the token movements as the transitions and the placement of tokens within the graph as the states. Consequently, the variables of a cTLA/c specification model the actual token placement on the activity, while its actions specify the flow of tokens between states. Flows may cross partition borders. According to the cTLA/c definition and because partitions are implemented by distributed components, flows moving between partitions are modeled by communication buffers, while states assigned to activity nodes are represented in cTLA/c by local variables. In order to define the semantics of activities using cTLA/c, we opted for an approach that directly uses the mechanisms of cTLA. We describe some activity element types as separate cTLA processes in (Khan & Heegaard, 2014), which helps to understand the semantics of the activity elements. Moreover, the production rules of cTLA actions for UML activities have been presented in (khan & Heegaard, 2014) to produce the system actions from the local process actions as a set of rules, so that each activity element can be defined separately. The concept of a UML deployment is also structural and describes a structure of the execution environment by identifying a system's physical layout. It also specifies which pieces of service components run on what pieces of physical nodes and how nodes are connected by communication paths. We use a specific tuple class as an additional invariant that is also a part of the style cTLA/c to model UML deployment, which is also described in (Khan & Heegaard, (2014).

cTLA to Specify UML STM Diagrams

Our modeling framework use UML STM diagrams to capture failure and recovery events of the system components. TLA is a linear-time temporal logic that models the system behavior where the system behavior is realized by a set of considerably large number of state sequences $s_0, s_1, s_2, \dots$ (Lamport, 2017; Khan, Machida, Heegaard, & Trivedi, 2012b). Thus, the TLA formalism can define the state machine formally produced by our framework, which ultimately also models considerably long sequences of states, s_i, starting with an initial state, s_0. cTLA originated from TLA to offer more easily comprehensible formalisms and propose a suppler composition of specifications 8. A state transition system is defined by the body of a cTLA process type. One cTLA process represents one state machine that mentions a set of TLA state sequences with the help of variables, actions, and events. The detailed formalism is defined in (Khan, Machida, Heegaard, & Trivedi, 2012b).

In addition, the formalization of a UML model using a cTLA process thus helps to describe the mapping process to show the correspondence with the analytical model. UML activities are based on Petri Nets and as such describe a state transition system. As an analytical model, our framework produces a SPN and SRN. The cTLA can define the state transitions formally produced by our framework, which ultimately also model considerably long sequences of states, s_i, starting with an initial state, s_0. Thus, we consider a mapping approach that directly shows the mapping of the cTLA-specified UML model into SRN model. This correspondence in turn ensures that the mapping of the cTLA-specified UML model and Petri nets fit together, and the UML correctly transforms to the analytical model. The detailed mapping process demonstrating the correspondence between the cTLA-specified UML and SRN is described in Khan & Heegaard, 2014.

Model Verification

Model verification is an integral part while deriving the tool-based support of the developed framework. This verification is also necessary to precisely and correctly represent the model specifications. Because temporal logic is applied to define the UML model specification, we can use the model checker TLC to verify the specification Khan, 2014. The external model checker TLC is invoked by Arctis from the command line, invisible to the user. To analyze building blocks and complete systems, the Arctis editor constantly checks the model for a number of syntactic constraints. For a more thorough analysis of the behavior, Arctis employs the model checker TLC based on the TLA. Figure 6 outlines this process: When a building block is complete and syntactically correct, Arctis transforms the UML activity into TLA+, the language for TLA, and initializes the model checker TLC. TLC can verify a specification for various temporal properties that are stated as theorems. For each activity, a set of theorems is automatically generated, which claims certain properties to be maintained by activities in general. Examples of these properties include the correct use of building blocks within the activity such that the activity itself satisfies a certain externally visible behavior; each call behavior action representing an instance of a building block, where a call behavior action declares a corresponding pin; each building block that has one activity partition for each collaboration role; state sequences that are connected through appropriate pins during the composition of building block activity, etc. When TLC detects that a theorem is violated, it produces an error trace displaying the state sequence that leads to the violation. If the model specification does not violate any theorems, the verification ends successfully. Otherwise, an error will be reported by the TLC in textual format, and a number of diagnoses will be considered based on the error trace. The detailed model verification process has been defined in (Kraemer, Slåtten, & Herrmann, 2007).

MODEL TRANSLATION

This section highlights the rules for the model translation from various UML models into SRN models. Since all the models will be translated into the SRN model, we will give a brief introduction about SRN model. SRN is based on the Generalized Stochastic Petri Net (GSPN) (Trivedi & Sahner, 2002) and extends them further by introducing prominent extensions such as guard function, reward function, and marking dependent firing rate (Trivedi & Sahner, 2002). A guard function is assigned to a transition. It specifies the condition to enable or disable a transition and can use the entire state of the net rather than just the number of tokens in places (Trivedi & Sahner, 2002). Reward function defines the reward rate for each tangible marking of Petri Net based on which various quantitative measures can be done in the Net level. Marking dependent firing rate allows using the number of tokens in a chosen place multiplied by the basic rate of the transition. SRN model has the following elements: Finite set of the place (drawn as circles), Finite set of the transition defined as either a timed transition (drawn as thick transparent bar) or an immediate transition (drawn as thick black bar), set of the arc connecting the place and transition, multiplicity associated with the arc, and marking that denotes the number of token in each place.

Before introducing the model translation rules, different types of collaboration roles as reusable basic building blocks are demonstrated with the corresponding SRN model in Table 1 that can be utilized to form the collaborative building blocks. The rules are the following:

Rule 1: The SRN model of a collaboration (Figure 7), where collaboration connects only two collaboration roles, is formed by combining the basic building blocks type 2 and type 3 from Table 1. Transition t in the SRN model is only realized by the overhead cost if service components A and B deploy on the same physical node as in this case, communication cost = 0, otherwise t is realized by both the communication and overhead cost.

In the same way, SRN model of the collaboration can be demonstrated where the starting of the execution of the SRN model of collaboration role A depends on the token received from the external source.

Rule 2: For a composite structure, when a collaboration role A connects with n collaboration roles by n collaborations like a star graph (where n > 1) where each collaboration connects only two collaboration roles, the SRN model is formed by combining the basic building block of Table I which is shown in Figure 8. In the first diagram of Figure 8, if component A contains its own token, equivalent SRN model of the collaboration role A will be formed using

Figure 6. Model Checker TLC

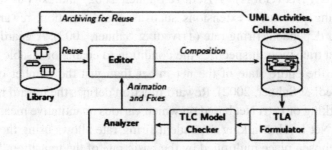

Table 1. Specification of reusable unites and corresponding SRN model

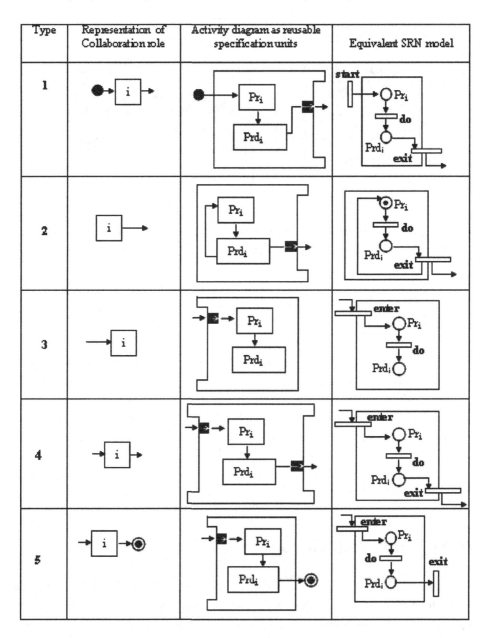

Figure 7. SRN Model Transformation

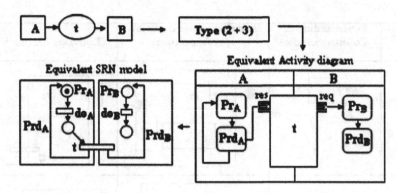

basic building block type 1 from Table 1. The same applies to the component *B* and *C* in the second diagram in Figure 8.

STM can be translated into SRN model by converting each state into place and each transition into a timed transition with input/output arcs which is reflected in the transformation Rule 3.

Rule 3: Rule 3 demonstrates the equivalent SRN model of the STM of hardware and software components which are shown in the Figure 9. The SRN model for hardware component is shown in Figure 9(b). A token in the place P_{run} represents the active hardware component and a token in P_{stb} represents a hot standby hardware component. When the transition T_{fail} fires, the token in P_{run} is removed and the transition T_{swt} is enabled. By the use of T_{swt}, which represents the failover, hot standby hardware component becomes an active component.

MODEL SYNCHRONIZATION

The model synchronization is achieved hierarchically which is illustrated in Figure 10. Performance SRN is dependent on the dependability SRN. Transitions in dependability SRN may change the behavior of the performance SRN. Moreover, transitions in the SRN model for the software process also depend on the transitions in the SRN model of the hardware component. These dependencies in the SRN models are handled through model synchronization by incorporating guard functions (Trivedi & Sahner, 2002). The model synchronization is described below:

Figure 8. SRN model transformation

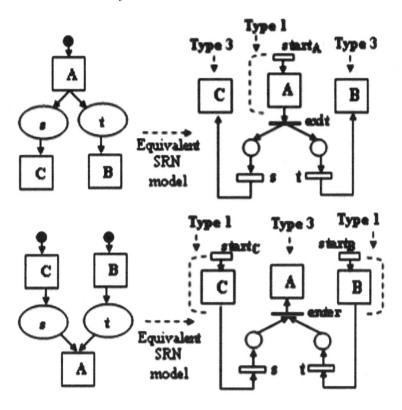

Synchronization Between the Dependability SRN Models in the Dependability Modeling Layer

SRN model for the software process (Figure 9(a)) is expanded by incorporating one additional place P_{hf}, three immediate transitions $t_{hf, thsfl, t_{hfr}}$, and one timed transition T_{recv} which helps to synchronize the transitions in the SRN model of the software process with the SRN model of the hardware component. The expanded SRN model (Figure 11(a)) is associated with four additional arcs such as $(P_{sfail} \times t_{hsfl}) \cup (t_{hsfl} \times P_{hf})$, $(P_{srec} \times t_{hfr}) \cup (t_{hfr} \times P_{hf})$, $(P_{srun} \times t_{hf}) \cup (t_{hf} \times P_{hf})$ and $(P_{hf} \times T_{recv}) \cup (T_{recv} \times P_{srun})$. The immediate transitions $t_{hf}, t_{hsfl}, t_{hfr}$ will be enabled only when the hardware node (in Fig. 11 (b)) fails as failure of hardware node stops the operation of the software process. The timed transition T_{recv} will be enabled only when the hardware node will again start working after being recovered from failure. Four guard functions $g_1, g_2,$ g_3, g_4 allow the four additional transitions $t_{hf}, t_{hsfl,} t_{hfr,}$ and T_{recv} of software process to work consistently with the change of states of the hardware node. The guard functions definitions are given in the Table 2.

Figure 9. SRN Model Transformation of UML STM Diagrams

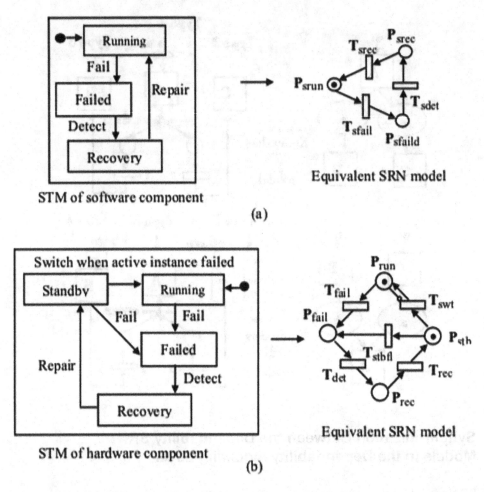

(a)

(b)

Figure 10. Model Synchronization Hierarchy

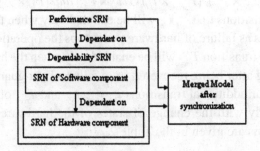

Synchronization Between the Dependability SRN and Performance SRN

In order to synchronize the collaboration role activity, performance SRN model is expanded by incorporating one additional place P_{fl} and one immediate transition f_A shown in Figure 12. When collaboration role "A" starts execution, a checking will be performed to examine whether both software and hardware components are running or not. If both the components work fine the timed transition do_A will be fired which represents the continuation of the execution of the collaboration role A. But if software resp. hardware components fail the immediate transition f_A will be fired which represents the quitting of the operation of collaboration role A. Guard function gr_A allows the immediate transition f_A to work consistently with the change of states of the software and hardware components.

Performance SRN model of parallel execution of collaboration roles are expanded by incorporating one additional place P_{fl} and immediate transitions f_{BC}, w_{BC} shown in Figure 12. In our discussion, during the synchronization of the parallel processes it needs to ensure that failure of one process eventually stops providing service. This could be achieved by firing of the immediate transition f_{BC}. If software resp. hardware components (Figure 11) fail immediate transition f_{BC} will be fired which symbolizes the quitting of the operation of both parallel processes B and C rather than stopping either process B or C, thus postponing the execution of the service. Stopping only either the process B or C will result in inconsistent execution of the system SRN and

Table 2. Guard functions

Function	Definition
g_1, g_2, g_3	if (# P_{run} = = 0) 1 else 0
g_4	if (# P_{run} = = 1) 1 else 0

Figure 11.

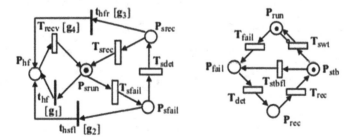

produce erroneous result. If both software and hardware components work fine the timed transition w_{BC} will be fired to continue the execution of parallel processes B and C. Guard functions gr_{BC}, grw_{BC} allow the immediate transition f_{BC}, w_{BC} to work consistently with the states change of the software and hardware components. The guard function definitions are shown in the Table 3.

HIERARCHICAL MODEL FOR MTTF CALCULATION

System is composed of different types of hardware devices such as CPU, memory, storage device, cooler. Hence, to model the failure behavior of a hardware node, we need to consider failure behavior of all the devices. But it is very demanding and not efficient with respect to execution time to consider behavior of all the hardware components during the SRN model generation. SRN model becomes very cumbersome and inefficient to execute. In order to solve the problem, we evaluate the mean time to failure (MTTF) of system using the hierarchical model in which a fault tree is used to represent the MTTF of the system by considering MTTF of

Table 3. Guard function

Function	Definition
gr_A, gr_{BC}	if (# P_{srun} = = 0) 1 else 0
grw_{BC}	if (# P_{srun} = = 1) 1 else 0

Figure 12. Model Synchronization

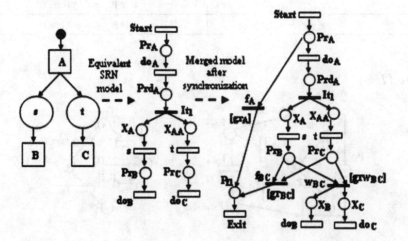

every hardware component available in the system. We consider this MTTF of the system in our dependability SRN model for hardware components (Figure 9(b)) rather than considering failure behavior of all the hardware components separately. The below Figure 13 introduces one example scenario of capturing failure behavior of the hardware components using fault tree where system is composed of different hardware devices such as one CPU, two memory interfaces, one storage device, and one cooler. The system will work when CPU, one of the memory interfaces, storage device, and cooler will run. Failure of both memory interfaces or failure of either CPU or storage device or cooler will result in the system unavailability.

CASE STUDY

As a representative example, we consider a scenario dealing with heuristically clustering of modules and assignment of clusters to nodes (Efe, 1982). This scenario is sufficiently complex to show the applicability of our performability framework. The problem is defined in our approach as collaboration of $E = 10$ service components or collaboration roles (labeled $C_1 \ldots C_{10}$) to be deployed and $K = 14$ collaborations between them illustrated in Figure 14. We consider three types of requirements in this specification. Besides the execution cost, communication cost, and cost for running background process, we have a restriction on components C_2, C_7, C_9 regarding their location. They must be bound to nodes n_2, n_1, n_3 respectively. In this scenario, new service is generated by integrating and combining the existing service components that will be delivered conveniently by the system. For example, one new service is composed of the service components C_1, C_2, C_4, C_5, C_7 shown in Figure 14 as

Figure 13. MTTF Calculation Scenario

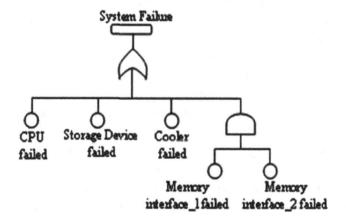

thick dashed line. The internal behavior of the collaboration K_i is realized by the call behavior actions through the UML activity diagram already demonstrated in Figure 3(b). The composition of the collaboration role C_i of the delivered service is demonstrated in Figure 15. The initial node ($\bullet$) indicates the starting of the activity. After being activated, each participant starts its processing which is mentioned by call behavior action Pr_i (Processing of the ith service component). Completions of the processing by the participants are mentioned by the call behavior action Prd_i (Processing done of the ith service component). The activity starts from the component C_7 where the semantics of the activity is realized by the token flow. After completion of the processing of the component C_7, the response is divided into two flows which are shown by the fork node f_7. The flows are activated towards component C_1 and C_4. After getting the response from the component C_1, processing of the components C_2 starts. The response and request are mentioned by the streaming pin *res* and *req*. The processing of the component C_5 starts after getting the responses from both component C_4 and C_2 which is realized by the join node j_5. After completion of the processing of component C_5, the activity terminates which is mentioned by the end node ($\bullet$).

In this example, the target environment consists of $N = 3$ identical, interconnected nodes with no failure of network link, with a single provided property, namely processing power, and with infinite communication capacities shown in Figure 16.

In order to annotate the UML diagrams in Figure 15 and 16, we use the stereotypes *<<saStep>> <<computingResource>>, <<scheduler>>* and the tagged values *execTime, deadline* and *schedPolicy* which are already explained in section 5. Collaboration K_i (Figure 17) is associated with two instances of *deadline* as collaborations in example scenario are associated with two kinds of cost: communication cost and cost for running background process (BP). In order to annotate the STM UML of the software process (shown in Figure 18), we use the stereotype *<<QoSDimension>>, <<transition>>* and attributes *mean-time-between-failures, mean-time-between-failure-detect* and *mean-time-to-repair* OMG, 2004. Annotation of the STM of hardware component can be demonstrated in the same way as STM of software process.

By considering the specification of reusable collaborative building blocks, deployment mapping, and the model transformation rules the corresponding SRN model of our example scenario is illustrated in Figure 19. In our discussion, we consider M/M/1/n queuing system so that at a time at most n jobs are available in the system (Trivedi, 2001). For generating the SRN model, firstly, we consider the starting node ($\bullet$). According to rule 1, it is represented by timed transition (denoted as start) and the arc connects to place Pr_7 (states of component C_7). When a token is deposited in place Pr_7, immediately a checking is done about the availability of both software and hardware components by inspecting the corresponding SRN models shown in Figure 11. The availability of software and hardware components allow the firing of timed transition t_7 that mentions the continuation of the further execution. Otherwise, immediate transition f_7 will be fired mentioning the ending of the further execution because of the failure of software resp. hardware components. The enabling of immediate transition f_7 is realized by the guard function gr_7. After the completion of the state transition from Pr_7 to Prd_7 (states of component C_7) the flow is divided into 2 branches (denoted by the immediate transition It_1) according to model transformation rule 2 (Figure 8). The token passes to place Pr_1 (states of component C_1) and Pr_4 (states of component C_4) after the firing of transitions K_7 and K_8. According to rule 1, collaboration K_8 is realized only by overhead cost as C_4 and C_7 deploy on the same processor node n_1 (Table IV). The collaboration K_7 is realized both by the communication cost and overhead cost as C_1 and C_7 deploy on the different physical nodes n_3 and n_1 (Table IV). When a token is deposited into place Pr_1 and Pr_4 a checking is done about the availability of both software and hardware components by inspecting the corresponding dependability SRN models illustrated in Figure 11. The availability of software and hardware components allows the firing of immediate transition w_{14} which eventually enables the firing of timed transition t_1 mentioning the continuation of the further execution. The enabling of immediate transition w_{14} is realized by the guard function grw_{14}. Otherwise, immediate transition f_{14} will be fired mentioning the ending of the further execution because of failure of software resp. hardware components. The enabling of immediate transition f_{14} is realized by the guard function gr_{14}. After the completion of the state transition from Pr_1 to Prd_1 (states of component C_1) the token is passed to Pr_2 (states of component C_2) according to rule 1, where timed transition K_5 is realized both by the communication and overhead cost. When a token is deposited into place Pr_2 a checking is done about the availability of both software and hardware components by inspecting the corresponding dependability SRN models shown in Figure 11. The availability of software and hardware components allows the firing of the immediate transition w_{24} which eventually enables the firing of timed transition t_2 and t_4 mentioning the continuation of the further execution. The enabling of immediate transition w_{24} is realized by the guard function grw_{24}. Otherwise, immediate transition f_{24} guided by guard function gr_{24} will be fired mentioning the ending of the further execution because of the failure of software resp. hardware components. Afterwards, the merging of the result is realized by the immediate transition It_2 following the firing of transitions K_2 and K_4. Collaboration K_2 is realized both by the overhead cost and communication cost as C_4 and C_5 deploy on the different processor nodes n_1 and n_2 (Table IV). K_4 is replaced by the timed transition which is realized by the overhead cost as C_2 and C_5 deploy on the same node n_2 (Table IV). When a token is deposited in place Pr_5 (state of component C_5), immediately, a checking is done about the availability of both software and hardware components by inspecting the corresponding SRN models illustrated in Figure 11. The availability of software and hardware components allow the firing of timed transition t_5 mentioning the continuation of the further execution. Otherwise, immediate transition f_5 will be fired mentioning the ending of the further execution because of the failure of software resp. hardware components and the ending of the execution of the SRN model is realized by the timed transition $Exit_2$. The enabling of immediate transition f_5 is realized by the guard function gr_5. After the completion of the state transition from Pr_5 to Prd_5 (states of component C_5) the ending of the execution of the SRN model is realized by the timed transition $Exit_1$. The definitions of guard functions gr_7, grw_{14}, gr_{14}, grw_{24}, gr_{24}, and gr_5 are mentioned in Table 4, which is dependent on the execution of the SRN model of the corresponding software and hardware instances illustrated in Figure 11.

We use SHARPE to execute the synchronized SRN model and calculate the system's throughput and job success probability against failure rate of system components. Graphs presented in Figure 20 show the throughput and job success probability of the system against the changing of the failure rate (sec^{-1}) of hardware and software components in the system.

CONCLUSION AND FUTURE WORK

We presented a novel approach for model based performability evaluation of a distributed system. The approach spans from the demonstration of the system's dynamics using UML diagram as reusable building blocks to efficient deployment of service components in a distributed manner focusing on the QoS requirements. The main advantage of using the reusable software components allows the cooperation among several software components to be reused within one self-contained, encapsulated building block. Moreover, reusability thus assists in creating distributed software systems from existing software components rather than developing the system from scratch which in turn facilitates the improvement of productivity and quality in accordance with the reduction in time and cost. We put emphasis to establish some important concerns relating to the specification and solution of performability models emphasizing the analysis of the system's dynamics. We design the framework in a

Table 4. Guard functions

Function	Definition
gr_7, gr_{14}, gr_{24}, gr_5	if (# P$_{srun}$ = = 0) 1 else 0
grw_{14}, grw_{24}	if (# P$_{srun}$ = = 1) 1 else 0

Figure 14. Service Composition Scenario

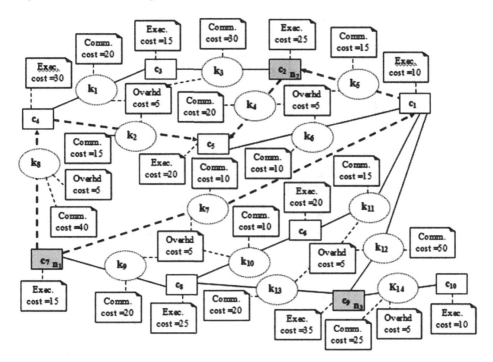

hierarchical and modular way which has the advantage of introducing any modification or adjustment at a specific layer in a particular sub-model rather than in the combined model. Among the important issues that come up in our development are flexibility of capturing system's dynamics using reusable specification of building blocks, ease of understanding the intricacy of combined model generation, and evaluating from the specification by proposing model transformation. However, our eventual goal is to develop support for runtime redeployment of components, this way keeping the service within an allowed region of parameters defined by the requirements. As a result, we can show with our framework that our logic will be a prominent candidate for a robust and adaptive service execution platform. The special property of SRN model like guard function keeps the performability model simpler by applying logical conditions that can be expressed graphically using input and inhibitor arcs which are easily realized by the following semantics: a logical "AND" for input arcs (all the input conditions must be satisfied), a logical "OR" for inhibitor arcs (any inhibitor condition is sufficient to disable the transition) (Trivedi & Sahner, 2002). However, the size of the underlying reachability set for generating SRN models is major limitation for large and complex systems. Further work includes the tackling the state explosion problems of reachability marking for large distributed systems.

Figure 15. UML Activity Diagram for the Defined Service

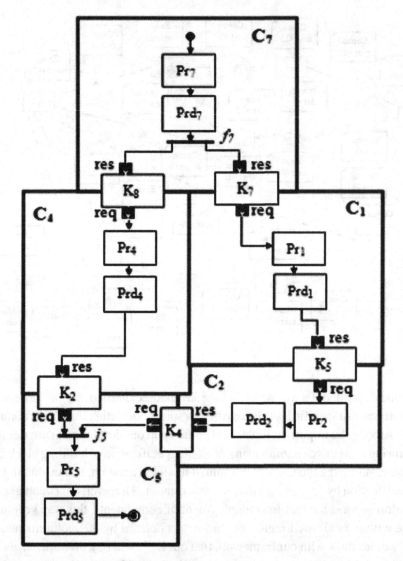

In addition, developing GUI editor is another future direction to generate UML deployment and STM and to incorporate performability related parameters. The plug-ins can be integrated into the Arctis which will provide the automated and incremental model checking while conducting model transformation.

Figure 16. UML Deployment Diagram

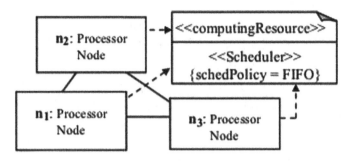

Figure 17. Annotated UML Model

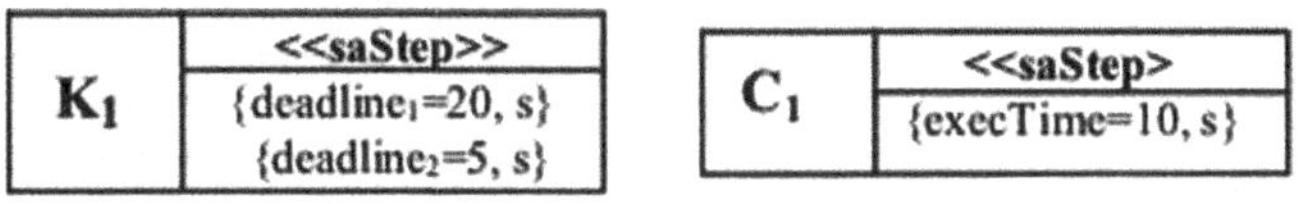

K_1	<<saStep>> {deadline$_1$=20, s} {deadline$_2$=5, s}	C_1	<<saStep> {execTime=10, s}

Figure 18. Annotate UML STM Diagrams

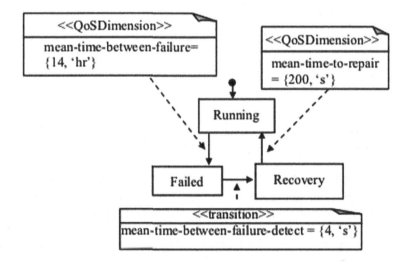

Figure 19. SRN Model of Our Defined Service

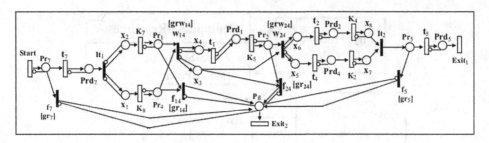

Figure 20. Results

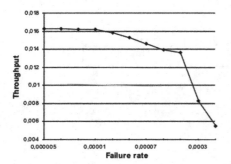 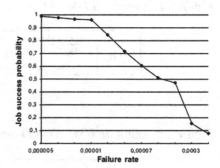

REFERENCES

Bracchi, P., & Cukic, B. (2004). Performability modeling of mobile software systems. In *Proceedings of the International Symposium on Software Reliability Engineering* (pp. 77-84). Bretange, France: IEEE. 10.1109/ISSRE.2004.27

Ciardo, G., Muppala, J., & Trivedi, K. S. (1992). Analyzing concurrent and fault-tolerant software using stochastic reward net. *Journal of Parallel and Distributed Computing, 15*(3), 255–269. doi:10.1016/0743-7315(92)90007-A

Csorba, M., Heegaard, P., & Hermann, P. (2008). Cost-Efficient deployment of collaborating components. In *Proceedings of the 8th IFIP WG 6.1 International conference on Distributed applications and interoperable systems* (pp. 253-268). Oslo, Norway: Springer. 10.1007/978-3-540-68642-2_20

Deri, G., & Varro. (2008). Model Driven Performability Analysis of Service Configurations with Reliable Messaging. In *Proceedings of the Model-Driven Web Engineering*, (pp. 61-75). Toulouse, France: CEUR.

Efe, K. (1982). Heuristic models of task assignment scheduling in distributed systems. *Computer.*

Herrmann, P., & Krumm, H. (2000). A framework for modeling transfer protocols. *Computer Networks, 34*(2), 317–337. doi:10.1016/S1389-1286(00)00089-X

Khan, R. H. (2014). *Performance and Performability Modeling Framework Considering Management of Service Components Deployment* (PhD Thesis). Norwegian University of Science and Technology, Trondheim, Norway.

Khan, R. H., & Heegaard, P. (2011). A Performance modeling framework incorporating cost efficient deployment of multiple collaborating components. In *Proceedings of the International Conference Software Engineering and Computer Systems* (pp. 31-45). Pahang, Malaysia: Springer. 10.1007/978-3-642-22170-5_3

Khan, R. H., & Heegaard, P. (2014). Software Performance evaluation utilizing UML Specification and SRN model and their formal representation. *Journal of Software, 10*(5), 499–523. doi:10.17706/jsw.10.5.499-523

Khan, R. H., Machida, F., & Heegaard, P., & Trivedi, K. S. (2012b). From UML to SRN: A performability modeling framework considering service components deployment. *International Journal on Advances in Networks and Services, 5*(3&4), 346–366.

Khan, R. H., Machida, F., Heegaard, P., & Trivedi, K. S. (2012a). From UML to SRN: A performability modeling framework considering service components deployment. In *Proceeding of the International Conference on Network and Services* (pp. 118-127). St. Marteen: IARIA.

Kraemer, F. A., Slåtten, V., & Herrmann, P. (2007). Engineering Support for UML Activities by Automated Model-Checking – An Example. In *Proceedings of the 4th International Workshop on Rapid Integration of Software Engineering Techniques*, (pp. 51-66). Academic Press.

Lamport. (2017). *Specifying Systems.* Addison-Wesley.

OMG. (2008). *UML Profile for Modeling Quality of Service and Fault Tolerance Characteristics and Mechanism.* Version 1.1. In Object management group. Retrieved January 2019, from https://pdfs.semanticscholar.org/e4bc/ffb49b6bd96d8df35173 704292866e0623eb.pdf

OMG. (2017). *OMG Unified Modeling Language™ (OMG UML) Superstructure.* Version 2.5. In Object management group. Retrieved January 2019 from http:// www.omg.org/spec/UML/2.2/ Superstructure/PDF/

OMG. (2019). *UML Profile for MARTE: Modeling and analysis of real-time embedded systems*. Version 1.2. In Object management group. Retrieved January 2019, from http://www.omg.org/omgmarte/ Documents/ Specifications/08-06-09.pdf

Sato, N., & Trivedi, K. S. (2007). Stochastic Modeling of Composite Web Services for Closed-Form Analysis of Their Performance and Reliability Bottlenecks. In *Proceedings of the International Conference on Service-Oriented Computing*, (pp. 107-118). Vienna, Austria: Springer. 10.1007/978-3-540-74974-5_9

Silva, E., & Gali, H. (1992). Performability analysis of computer systems: From model specification to solution. *Performance Evaluation*, *14*(3-4), 157–196. doi:10.1016/0166-5316(92)90003-Y

Slåtten, V. (2007). *Model Checking Collaborative Service Specifications in TLA with TLC* (Project Thesis). Norwegian University of Science and Technology, Trondheim, Norway.

Trivedi, K. S. (2001). Probability and Statistics with Reliability, Queuing and Computer Science application. Wiley-Interscience Publication.

Trivedi, K. S., & Sahner, R. (2002). *Symbolic Hierarchical Automated Reliability / Performance Evaluator (SHARPE)*. Department of Electrical and Computer Engineering, Duke University.

Wet, N. D., & Kritzinger, P. (2011). *Towards Model-Based Communication Protocol Performability Analysis with UML 2.0*. Retrieved January 2011, from http://pubs.cs.uct.ac.za/archive/00000150/ 01/ No_10

KEY TERMS AND DEFINITIONS

Analytic Approach: Analytic approach such as Markov model and petri net are used to evaluate the non-functional properties of the systems.

Collaborative Building Block: Collaborative building block is the main specification unit defined by UML collaboration and activity which captures the interaction and detail behavior between the software components.

Distributed System: A distributed system consists of multiple autonomous unit that communicate through a network in order to achieve a common goal.

Formal Method: It provides mathematical modeling through which a system can be specified and verified.

Model Transformation: To transform the UML model into analytical model (e.g., Markov, SPN, SRN) is defined as model transformation.

Performability: Combined approach evaluate the performance and dependability parameters of the system.

Reusable Building Block: Collaborative building block is called as reusable building block as it is archived in a library for reusing purpose.

Service: A software system to achieve specific goals in a computing environment.

Temporal Logic: Temporal logic provides the symbols and rules to reason about a proposition in respect to time.

UML Profile: Profile in UML is defined using stereotypes, tag definitions, and constraints that are applied to specific model elements, such as classes, attributes, operations, and activities.

Chapter 8
Sample Size Equation of Mechanical System for Parametric Accelerated Life Testing

Seongwoo Woo
Ababa Science and Technology University, Ethiopia

ABSTRACT

This chapter proposes how to decide the sample size. Due to cost and time, it is unfeasible to carry out the reliability testing of product with many samples. More studies on testing methods should explain how to decrease the sample size. To provide the accurate analysis for product reliability in the bathtub curve, Weibull distribution is chosen. Based on Weibayes method, we can derive the sample size equation from the product BX life. As the reliability of product is targeted and the limited sample size is given, we can determine the mission cycles from new sample size equation. Thus, by applying the accelerated loading to mechanical product, we can reveal the design failures of product and confirm if the reliability target is achieved. Comparing conventional sample size equation, this derivation of sample size equation is helpful to evaluate the lifetime of mechanical product and correct its design problems through parametric accelerated life testing (ALT). This equation also might be applicable to develop the test method of all sorts of structure in the civil engineering.

DOI: 10.4018/978-1-7998-1718-5.ch008

INTRODUCTION

To prevent the massive product recall due to the faulty designs from the marketplace, mechanical system in the product developing process is required to be tested with a proper reliability methodology for assessing its lifetime before launching product (See Figure 1). The reliability of a mechanical system can be defined as the ability of a system or module to function under stated conditions for a specified period of time (IEEE, 1990). It can be illustrated by the bathtub curve that consists of three parts (Klutke, Kiessler, & Wortman, 2015). There is initially a decreasing failure rate, a constant failure rate, and then an increasing failure rate. If product follows the bathtub curve, it will have difficulty in succeeding in the marketplace because product failures frequently occur in its lifetime. In the initial introduction of mechanical product, the methodology of reliability quantitative test should be established to identify the failures of new mechanical product and secure its reliability. However, these reliability current methodologies (O'Connor & Kleyner, 2012; Bertsche, 2010; Yang, 2007; Lewi, 1995; Saunders, 2007). fail to reproduce the design faults and still have been controversial because small samples among products are tested and product failures rarely occur in field.

Especially, identifying the product failures is not easy because the faulty designs (voids) of certain structures subjected to repetitive stresses suddenly provoke trouble in its lifetime. Furthermore, correcting them, we try to ensure the lifetime of a mechanical product at the R&D stage. To achieve these goals, we should develop new reliability methodology – parametric ALT. But there are some issues to be solved: 1) reliability test plans of multi-module product based on its reliability definition and block diagram, 2) acceleration method of mechanical system, 3) derivation of sample size equation combined with reliability target and accelerated factor, 4) accelerated life testing with reproduction of design problems and its action plans, and 5) evaluation of the final product design to ensure the reliability criteria are satisfied.

Among them we will discuss about deriving the sample size equation to apply the reliability testing. It might be included as the following terms: 1) confidence level, 2) the allowed number of failure, 3) test condition, 4) test period, 5) the acceleration factor, and 6) reliability target, etc. However, there was no clear research about the sample size on the reliability quantitative test of product (Sandelowski, 1995; Guest, Bunce, & Johnson, 2006). For this reason, it is required to carry out study on determining optimal sample size for the reliability quantitative test of mechanical system. After acquiring failed ones with planned test, we can search out product failures and improve the faulty designs of mechanical system with actions plans (Woo, O'Neal, & Pecht, 2009a; 2009b; 2010).

To determining sample size, the Weibayes method for reliability data of product is well-known method (Weibull, 1951; Martz & Waller, 1982). Because of mathematical

Figure 1. Mechanical product and its failure

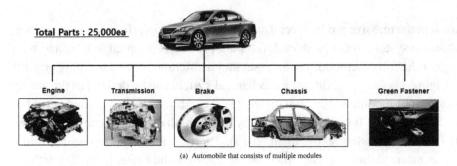

(a) Automobile that consists of multiple modules

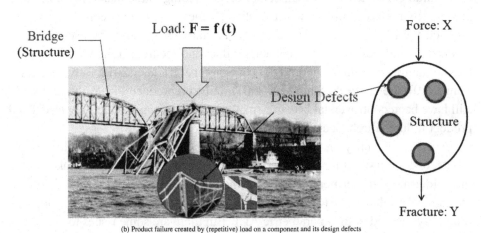

(b) Product failure created by (repetitive) load on a component and its design defects

complexity including statistical concepts, it is not easy to directly derive the sample size. For example, to estimate the characteristic life η in a confidence level, we have to separate two cases of type II censoring: 1) failures ($r \geq 1$) and 2) no failure ($r = 0$). To conveniently calculate the sample size from some software or by hand, we also develop the simplified equation under statistical assumptions. But from the final approximated sample size equation it is difficult to understand the statistical background of concealed algorithm. To utilize the test method by this algorithm, we need to clearly describe the induced process that might decide sample size and test time (or mission cycles).

To perform lifetime test, we need to target the durability index of a mechanical product. The index might be defined as BX life, which is the time at which X percent of items in a population will have failed. For example, if B20 life of a mechanical product is 10 years, 20 percent of the population will have failed during 10 years of operation time. The benefit of 'BX life Y years' as product reliability is to simultaneously represent the cumulative failure rate of a product and its lifetime corresponding to market requirement. On the other hands, the Mean Time To Failure (MTTF) as the inverse of the failure rate cannot no longer stand for the product lifetime, because MTTF approximately signifies B60 life and the index is too long to reach the product target. Consequently, when mechanical engineer targets product lifetime, BX life is more adequate than that of MTTF.

This study is to suggest two sample size equations: 1) general sample size equation and 2) approximated sample size equation that will be the basis of the reliability methodology of mechanical system. First of all, we utilize the Weibayes method that can be defined as Weibull Analysis with a given shape parameter. Secondly after the conversion from characteristic life to BX life we will derive the general sample size equation. Because the general sample size equation has difficult in conveniently calculating the mission cycle by hands, we will derive the approximated sample size equation under 60% confidence levels. Consequently, if the reliability target is assigned, we can obtain the mission cycles of newly designed product and perform its reliability testing from the sample size equation.

DETERMINATION OF SAMPLE SIZE

Weibull Distribution for Analyzing Product Lifetime

Weibull Distribution Function

If a random variable T is represented to the lifetime of product, the Weibull Cumulative Distribution Function (CDF) of T, denoted $F(t)$, may be defined as all individuals having an $T \leq t$, divided by the total number of individuals. This function also gives the probability P of randomly choosing an individual having a value of T equal to or less than t, and thus we have

$$P\left(T \leq t\right) = F\left(t\right) \tag{1}$$

If module in product follows Weibull distribution, the accumulated failure rate, $F(t)$, is expressed as:

$$F\left(t\right) = 1 - e^{-\left(\frac{t}{\eta}\right)^{\beta}} \tag{2}$$

Equation (2) gives the appropriate mathematical expression for the principle of the weakest link in the product, where the occurrence of an failure in any part of an product may be said to have occurred in the product as a whole, e.g., the phenomena of yield limits, static or dynamic strengths of product, life of electric bulbs, even death of man.

As seen in Figure 2, the summation of $F(t)$ and $R(t)$ is always one as follows:

$$F\left(t\right) + R\left(t\right) = 1, \text{ for } t \geq 0.$$ (3)

So the product reliability is expressed as follow:

$$R\left(t\right) = e^{-\left(\frac{t}{\eta}\right)^{\beta}}$$ (4)

where t is time, η is characteristic life, and β is shape parameter (the intensity of the wear-out failure).

Figure 2. Relationship between F(t) and R(t) at time t_1 that might be described in bathtub

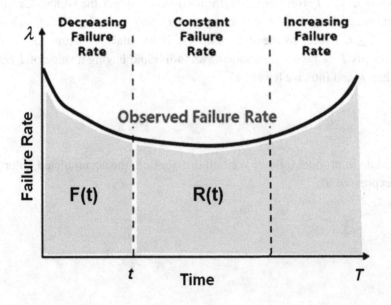

Weibayes Method

Failures can be classified into three categories, which are infant mortality, random failure, and wear-out failure in the bathtub. These categories can be explained according to shape parameter of the Weibull distribution. For estimating lifetime, shape parameter might be greater than one. The Weibayes method can be defined as Weibull Analysis with a given shape parameter. If the shape parameter is assumed from marketplace or experimental data, characteristic life from Weibayes can be derived from using Maximum Likelihood Estimate (MLE) as follow (Guest, Bunce, & Johnson, 2006):

$$\eta_{MLE}^{\beta} = \sum_{i=1}^{n} t_i^{\beta} / r \qquad (5)$$

where η_{MLE} is maximum likelihood estimate of the characteristic life, n is total number of sample, t_i is test duration for each sample, and r is the number of failures.

When the number of failure is $r \geq 1$ and confidence level is $100(1-\alpha)$, characteristic life, η_{α}, would be estimated as the following middle term, and changed into the last term using Equation (5),

$$\eta_{\alpha}^{\beta} = \frac{2r}{\chi_{\alpha}^2 (2r+2)} \cdot \eta_{MLE}^{\beta} = \frac{2}{\chi_{\alpha}^2 (2r+2)} \cdot \sum_{i=1}^{n} t_i^{\beta} \text{ for } r \geq 1 \qquad (6)$$

Presuming no failures are found, characteristic life, η_{α}. would be shown as follow:

$$\eta_{\alpha}^{\beta} = \frac{1}{\ln \dfrac{1}{\alpha}} \cdot \sum_{i=1}^{n} t_i^{\beta} \text{ for } r = 0 . \qquad (7)$$

Here, the first term, $\ln \dfrac{1}{\alpha}$, is mathematically equivalent to Chi-Squared value, $\dfrac{\chi_{\alpha}^2 (2)}{2}$, when p-value is α. Then, Equation (6) can be applied to all cases, or

$$\eta_\alpha^\beta = \frac{2}{\chi_\alpha^2(2r+2)} \cdot \sum_{i=1}^{n} t_i^\beta \text{ for } r \geq 0.$$ (8)

The Conversion From Characteristic Life to BX Life

In order to evaluate the product durability, the characteristic life in the Weibull distribution should be changed into the BX life. The characteristic life can be converted into BX life as follows:

$$R(t) = e^{-\left(\frac{L_{BX}}{\eta}\right)^\beta} = 1 - x.$$ (9)

$$L_{BX}^\beta = \left(\ln\frac{1}{1-x}\right) \cdot \eta^\beta.$$ (10)

where L_{BX} is BX life, and x is cumulative failure rate till lifetime ($x = X / 100$).

If characteristic life, η, in Equation (9) is substituted with the estimated characteristic life of p-value α, η_α, in Equation (7), we arrived at the following BX life equation:

$$L_{BX}^\beta = \frac{2}{\chi_\alpha^2(2r+2)} \cdot \left(\ln\frac{1}{1-x}\right) \cdot \sum_{i=1}^{n} t_i^\beta.$$ (11)

The following equation found in case of no failure is identical with Equation (11).

$$L_{BX}^\beta = \frac{\ln(1-x)}{\ln\alpha} \cdot \sum_{i=1}^{n} t_i^\beta = \frac{1}{\ln\frac{1}{\alpha}} \cdot \left(\ln\frac{1}{1-x}\right) \cdot \sum_{i=1}^{n} t_i^\beta.$$ (12)

Large Sample for Production

If the sample size is large, most items will operate till the test plan. That is,

$$\sum_{i=1}^{n} t_i^{\beta} \cong nh^{\beta}, \tag{13}$$

where h is the planned test time.

The estimated lifetime $\left(L_{BX}\right)$ drawn from the test should be longer than the target lifetime $\left(L_{BX}^{*}\right)$.

$$L_{BX}^{\beta} \cong \frac{2}{\chi_{\alpha}^{2}\left(2r+2\right)} \cdot \left(\ln\frac{1}{1-x}\right) \cdot nh^{\beta} \geq L_{BX}^{*\beta}. \tag{14}$$

Finally, we can find that sample size equation is expressed as follow:

$$n \geq \frac{\chi_{\alpha}^{2}\left(2r+2\right)}{2} \cdot \frac{1}{\left(\ln\frac{1}{1-x}\right)} \cdot \left(\frac{L_{BX}^{*}}{h}\right)^{\beta}. \tag{15}$$

Small Sample for R&D

In developing product, it is not available to acquire sufficient samples. Compared to anticipating failures in lifetime test, the allowed number of failures would not be small. That is,

$$\sum_{i=1}^{n} t_i^{\beta} = \sum_{i=1}^{r} t_i^{\beta} + \left(n-r\right)h^{\beta} \geq \left(n-r\right)h^{\beta}. \tag{16}$$

So BX life equation might be changed as follow:

$$L_{BX}^{\beta} \geq \frac{2}{\chi_{\alpha}^{2}\left(2r+2\right)} \cdot \left(\ln\frac{1}{1-x}\right) \cdot \left(n-r\right)h^{\beta} \geq L_{BX}^{*\beta}. \tag{17}$$

Rearrange Equation (17), we can obtain the sample size equation. So this equation adds the allowed number of failure from the previous Equation (15):

$$n \geq \frac{\chi_\alpha^2(2r+2)}{2} \cdot \frac{1}{\left(\ln\dfrac{1}{1-x}\right)} \cdot \left(\frac{L_{BX}^*}{h}\right)^\beta + r .$$ (18)

And we say that this will become the generalized sample size equation. When identified any failure of mechanical product, we can carry out the test of product lifetime.

APPLICATION

Approximated Equation for Sample Size

In the developing process of mechanical product, high confidence level would be preferred. But because of effectiveness including cost, test period, etc., common sense level of confidence will be sufficient, which reaches to around 60 percent. When the confidence level is 60%, the first term in Equation (18), $\dfrac{\chi_\alpha^2(2r+2)}{2}$, can be converted to $(r+1)$ (Table 1).

And if the cumulative failure rate, x, is sufficiently small, the denominator of the second term in Equation (18),

Table 1. Chi-square distribution in 60% (or 90%) confidence level

C.L=60%		$\frac{\chi_\alpha^2(2r+2)}{2}=(r+1)$		C.L=90%		$\frac{\chi_\alpha^2(2r+2)}{2}=2(r+1)$		
r	$1-\alpha$	$\frac{\chi_{0.4}^2(2r+2)}{2}$	$\frac{\chi_\alpha^2(2r+2)}{2}$ (a)	$1-\alpha$	$1-\alpha$	$\frac{\chi_{0.1}^2(2r+2)}{2}$ (b)	(b)/(a)	$\frac{\chi_\alpha^2(2r+2)}{2}$	$1-\alpha$
0	0.6	0.92	1	0.63	0.9	2.30	2.3	2	0.86
1	0.6	2.02	2	0.59	0.9	3.89	1.9	4	0.91
2	0.6	3.11	3	0.58	0.9	5.32	1.8	6	0.94
3	0.6	4.18	4	0.57	0.9	6.68	1.7	8	0.96

$$\ln \frac{1}{1-x} = \left(x + \frac{x^2}{2} + \frac{x^3}{3} + \cdots \right),$$

approximates to x. Then sample size equation can be simplified as follow:

$$n \geq (r+1) \cdot \frac{1}{x} \cdot \left(\frac{L_{BX}^*}{h} \right)^{\beta} + r. \tag{19}$$

Here when we introduce accelerated conditions, it will become approximated equation on sample size determination with common sense level of confidence, as follow:

$$n \geq (r+1) \cdot \frac{1}{x} \cdot \left(\frac{L_{BX}^*}{AF \cdot h_a} \right)^{\beta} + r, \tag{20}$$

where h_a is actual test hours and AF is acceleration factor.

Most of acceleration factors can be expressed as follows (Woo & O'Neal, 2010):

$$AF = \left(\frac{S}{S_0} \right)^n \cdot \exp\left[\frac{E_a}{k} \left(\frac{1}{T_0} - \frac{1}{T} \right) \right] = \left(\frac{e}{e_0} \right)^{\lambda} \cdot \exp\left[\frac{E_a}{k} \left(\frac{1}{T_0} - \frac{1}{T} \right) \right], \tag{21}$$

where S_0, e_0, T_0 are the stress, effort, and absolute temperature under normal operating conditions respectively, S, e, T are the stress, effort, and absolute temperature under accelerated conditions, n is the exponent of stress, E_a is activation energy, and k is the Boltzmann constant.

Because power is not easy to measure directly, it can be represented as two temporary variables – flow and effort in multiple physical domains. That is,

$$P = e(t) \cdot f(t) \tag{22}$$

Table 2 also shows the physical meanings of the variables in different domains.

Parametric ALT for Mechanical System – Robot Actuator

Robot Actuator

Parametric ALT is an accelerated life test for quantification using with parameters of the approximated equation and the AF equation; shape parameter (β), the allowed number of failures (r), the exponent of stress (n), and activation energy (E_a).

To applying them to mechanical system - robot, reliability is targeted to have B20 life years. Because robot consists of fifth module, reliability for actuator module might be B4 life 5 years by part count method. So the calculated lifetime of product would be 3,600 hours for 5years

$$\left(= 2hour \, / \, Day \cdot 360Day \, / \, Year \cdot 5Year\right).$$

Now consider a test plan for evaluating the mechatronic actuator that consists of DC motor, gear train and power MOSFET as shown in Figure 3.

Because the stresses of actuator come from torque, Equation (21) can be modified as Equation (23). As the normal conditions are at 0.10 kg·cm (100mA) and 20°C, accelerated conditions are elevated to 0.16 kg·cm (140mA) and 40°C. Acceleration factor can be calculated by using as follow:

Table 2. Energy flow in the multi-port physical system

Modules	Effort, $e(t)$	Flow, $f(t)$
Mechanical translation	Force, $F(t)$	Velocity, $V(t)$
Mechanical rotation	Torque, $\tau(t)$	Angular velocity, $\omega(t)$
Compressor, Pump	Pressure difference, $\Delta P(t)$	Volume flow rate, $Q(t)$
Electric	Voltage, $V(t)$	Current, $i(t)$
Thermal	Temperature, T	Entropy change rate, ds/dt
Chemical	Chemical potential, μ	Mole flow rate, dN/dt
Magnetic	Magneto-motive force, e_m	Magnetic flux, φ

Figure 3. The composition of actuator for test plan

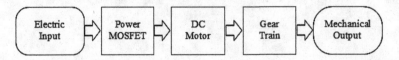

$$AF = \left(\frac{T}{T_0}\right)^{\lambda} \cdot \exp\left[\frac{E_a}{k}\left(\frac{1}{T_0} - \frac{1}{T}\right)\right] = 1030 \cong 10 \,. \tag{23}$$

where the activation energy is 0.55eV (typical reaction rate), exponent, λ, is 2 because the loss of electrical input power dissipates in conductor by I^2R (Hughes, 1993).

So by the acceleration factor the actual test time will be reduced to 360 hours (1/10 times). To assure the lifetime target B4 life 5 years, 25 samples might be tested in case of no failure allowed ($r=0$) because the last term of Equation (20) becomes one. We know that this sample size still is required to have too many numbers for the reliability testing of mechanical product. As test time increases, we can find the suitable solution. For instance, if the test time is extended to 1080 hours (360 hours×3 times) and shape parameter is three, the sample size will decrease by the cube of one-third ($=(1/3)^3$). Consequently, the required sample size will decrease to one.

Compared with those of Minitab, Table 3 shows the calculation results of sample size for mechanical system – robot actuator under several conditions. You should note that for Minitab there is no input term on accelerated condition (or accelerated factor) when sample size is calculated.

According to Table 3, when shape parameter is three, we know that one actuator should be tested for 1080 hours under conditions of 0.16 kg·cm (160 mA) and 40 ° . (temperature). If no failure happens in this test, the lifetime target B4 life 5 years would be guaranteed with common sense level of confidence. This test would be adequate to identify the failures of mechanical product subjected to repetitive mechanical load. In Table 4 we also suggest the comparison between the approximated equation and the generalized equation with 60 percent confidence level. As seen in this table, the approximated equation might be replaced with generalized equation.

Table 3. Sample size according to shape parameter and number of failure

Shape Parameter (β)	Number of Failure Allowed (r)	Test Time (h_a)	Sample Size by Equation (19)	Sample Size by Minitab
2	0	1,080h (45days)	3	3
2	1	1,080h (45days)	7	7
3	0	1,080h (45days)	1	1
3	1	1,080h (45days)	3	3

Consequently, the approximated equation is beneficial to calculate the sample size of mechanical system under a variety of accelerated conditions.

Printed Circuit Assembly (PCA)

To applying parametric ALT to PCA in mechanical system, reliability is targeted to have B4 life 5 years. So the calculated lifetime of product due to thermal shock would be 1,800 cycles for 5years

$$\left(= 1cycle \: / \: Day \cdot 360Day \: / \: Year \cdot 5Year\right).$$

Because the stresses of PCA come from system thermal shock, Equation (21) can be modified as Equation (24). As the thermal shock under normal conditions is at $\Delta T=10°C$, the thermal shock under accelerated conditions is elevated to $\Delta T=50°C$. Acceleration factor can be calculated by using as follow:

$$AF = \left(\frac{\Delta T}{\Delta T_0}\right)^{\lambda} = \left(\frac{50}{10}\right)^2 \cong 25 \tag{24}$$

where exponent, λ, is 2.

Table 4. Comparison between approximated equation and generalized equation with 60 percent confidence level

		Approximated Equation (Equation (19))	Generalized Equation (Equation (17))
		$\left(r+1\right) \cdot \dfrac{1}{x} \cdot \left(\dfrac{L_{BX}^*}{AF \cdot h_a}\right)^{\beta} + r$	$\dfrac{\chi_{0.4}^2\left(2r+2\right)}{2} \cdot \dfrac{1}{\left(\ln\dfrac{1}{1-x}\right)} \cdot \left(\dfrac{L_{BX}^*}{h}\right)^{\beta} + r$
2	0	$2.78 \cong 3$	$2.49 \cong 3$
2	1	$6.56 \cong 7$	$6.50 \cong 7$
3	0	$0.93 \cong 1$	$0.83 \cong 1$
3	1	$2.85 \cong 3$	$2.83 \cong 3$

So by the acceleration factor the actual test time will be reduced to 72 cycles (=1/25 times). To assure the lifetime target B4 life 5 years, 25 samples might be tested in case of no failure allowed (r=0) because the last term of Equation (20) becomes one. Because sample size is required to have many numbers for the reliability testing of mechanical product, we increase the test time. That is, the test time is extended to 260 cycles (=3.6 times). If shape parameter is two, the sample size will decrease to two (=(1/13)).

CONCLUSION

Based on the Weibayes method, the generalized equation has been derived for determining sample size. For application, we have approximated this sample size equation under 60% confidence levels. Comparing with the generalized sample equation, the approximated equation of sample size has relatively accurate results in practical range for the lifetime of mechanical product. If reliability is targeted, we obtain the mission cycles from the approximation equation. As any defective configurations in the product developing process are found, we correct them immediately before launching product. Therefore, the reliability test could be sufficiently applied to the mechanical system in the R&D.

REFERENCES

Abernethy, R. B. (2006). The New Weibull Handbook (5th ed.). Academic Press.

Bertsche, B. (2010). *Reliability in Automotive and Mechanical Engineering: Determination of Component and System Reliability*. Berlin: Springer.

Guest, G., Bunce, A., & Johnson, L. (2006). How many interviews are enough?: An experiment with data saturation and variability. *Field Methods, 18*(1), 59–82. doi:10.1177/1525822X05279903

Hughes, A. (1993). *Electric Motors and Drives*. Oxford, UK: Newness.

IEEE Standard Glossary of Software Engineering Terminology. (1990). *IEEE Std 610.12-1990*. New York, NY: Standards Coordinating Committee of the Computer Society of IEEE.

Klutke, G., Kiessler, P. C., & Wortman, M. A. (2015). A critical look at the bathtub curve. *IEEE Transactions on Reliability, 52*(1), 125–129. doi:10.1109/TR.2002.804492

Lewi, E. (1995). *Introduction to Reliability Engineering*. Wiley.

Martz, H. F., & Waller, R. A. (1982). *Bayesian Reliability Analysis*. New York: John Wiley & Sons, Inc.

O'Connor, P., & Kleyner, A. (2012). *Practical Reliability Engineering (5th ed.)*. Wiley & Sons.

Sandelowski, M. (1995). Sample size in qualitative research. *Research in Nursing & Health, 18*(2), 179–183. doi:10.1002/nur.4770180211 PMID:7899572

Saunders, S. (2007). *Reliability, Life Testing and the Prediction of Service Lives*. New York: Springer. doi:10.1007/978-0-387-48538-6

Weibull, W. (1951). A statistical distribution function of wide applicability. *Journal of Applied Mechanics, 18*, 293–297.

Woo, S., & O'Neal, D. (2010). Reliability design and case study of mechanical system like a hinge kit system in refrigerator subjected to repetitive stresses. *Engineering Failure Analysis, 99*, 319–329. doi:10.1016/j.engfailanal.2019.02.015

Woo, S., O'Neal, D., & Pecht, M. (2009a). Improving the reliability of a water dispenser lever in a refrigerator subjected to repetitive stresses. *Engineering Failure Analysis, 16*(5), 1597–1606. doi:10.1016/j.engfailanal.2008.10.017

Woo, S., O'Neal, D., & Pecht, M. (2009b). Design of a Hinge kit system in a Kimchi refrigerator receiving repetitive stresses. *Engineering Failure Analysis, 16*(5), 1655–1665. doi:10.1016/j.engfailanal.2008.11.010

Woo, S., O'Neal, D., & Pecht, M. (2010). Failure analysis and redesign of the evaporator tubing in a Kimchi refrigerator. *Engineering Failure Analysis, 17*(2), 369–379. doi:10.1016/j.engfailanal.2009.08.003

Yang, G. (2007). *Life Cycle Reliability Engineering*. Wiley. doi:10.1002/9780470117880

Chapter 9
Risk Factor in Agricultural Sector:
Prioritizing Indian Agricultural Risk Factor by MAUT Method

Suchismita Satapathy

https://orcid.org/0000-0002-4805-1793
KIIT University, India

ABSTRACT

Occupational safety is a big issue of discussion for agricultural workers. The methods of working in the field in extreme climate totally depends on the environmental factor. Due to change in weather conditions, prices at the time of harvest could drop, hired labour may not be available at peak times, machinery and equipment could break down when most needed, animals might die, and government policy can change overnight. All of these changes are examples of the risks that farmers face in managing their farm as a business. All of these risks affect their farm profitability. Heavy rains and drought could also damage or even wipe out crops. Another source of production risk is equipment. The most common sources of risk factor are weather, climate, diseases, natural disasters, and market and environmental factor shocks. Agricultural workers need sufficient precaution and safety measures at the time of field and machine work to minimize risk factor. So, in this chapter, an effort is taken to prioritize safety majors by MAUT method.

DOI: 10.4018/978-1-7998-1718-5.ch009

INTRODUCTION

The utilization of genetically modified crops and Organic farming will improve the fertility of the land and improve the crop production rate of Indian farmers. But still, the small and medium agricultural sector is very poor and neglected, and by following the traditional method of crop production. Risk and uncertainty are inherent to agriculture. The most common sources of risk factors are the weather, climate, diseases, natural disasters, the downturn of agricultural market and environmental factors. Some risks have become more severe in recent years due to climate change and the volatility of food prices. nominal farmers' livelihoods are especially vulnerable. They may have difficulty in finding and organization risk and fail to benefit from investment opportunities that could progress their farming businesses and strengthen household resilience. Farmers live with risk and make decisions every day that affect their farming operations. Due to changes in weather conditions prices at the time of harvest could drop; hired labor may not be available at peak times; machinery and equipment could break down when most needed; draught animals might die, and government policy can change overnight. All of these changes are examples of the risks that farmers face in managing their farm as a business. All of these risks affect their farm profitability. Heavy rains and Draught without rain could also damage or even wipe outcrops. Another source of production risk is the nonavailability of modern equipment.

The price of farm products is affected by the supply of a product, demand for the product, and the cost of production. The technology, assets, and labor or human factor is also a very important issue. Contacts and exposure with the chemicals, fertilizers), the exposure to soil, dust, the contamination due to bacteria, exposure to animals, cattle's, injury due to hand tools and musculoskeletal disorders are the most important injuries faced by farmers. Indian agricultural business sector is expected to be the most important driver of the Indian economy within a few years because of high investments for agricultural facilities, warehousing, and cold storage. The utilization of genetically modified crops and Organic farming will improve the fertility of the land and improve the crop production rate of Indian farmers. But still, the small and medium agricultural sector is very poor and neglected, and are following the traditional method of crop production. Due to the high cost of equipment unable to purchase and the conventional method of farming gives them many physical problems like lungs problem due to exposure to dust, and musculoskeletal disorders. Apart from this, Financial or economic risk plays an important part, when money is borrowed to finance the farming business. This risk can be caused by uncertainty about future interest rates, a lender's willingness and ability to continue to provide funds when needed, and the ability of the farmer to generate the income necessary for loan repayment. Human risk factor is the major risks to the farm business caused by illness or death and the personal situation of the farmer family. Accidents, illness, and death can disrupt farm performance. Due to high cost of equipment farmers unable to purchase modern equipments and the conventional method of farming gives them many physical problems like lungs problem due to exposure to dust, and muskulateral disorders.

Extreme weather conditions, the heavy workload during their working procedure gives them health and economic problems in farming. So to attain better efficiency of performance and to improve the productivity of the worldwide farmers in the agricultural sector it is essential to design the tools and equipment to make their work easy and risk-free but the design must be done keeping in consideration the farmer's capability and limits. The tools and equipments design should be able to provide more human comfort, of good quality, more output focused and reduce the musculoskeletal injury.By avoiding all these factors risk factors will reduce.

LITERATURE REVIEW

The literature on agricultural injuries and accidents in India is limited to a few. Several authors and researchers have reported the rate of agricultural accidents as higher than the industrial sectors. The lack of infrastructural facility, medical facility, training and rare use & maintenance of machines as well as the lack of defining the work according to gender or age are the causes of agricultural injuries (Knapp, 1965, 1966). Mohan and Patel (1992) have reported 576 agricultural related injuries in 1 year in 9 villages of Haryana in India. 87% injuries were reported as minor, 11% as moderate and remaining as severe. It is found that most of the injuries were due to hand tools, i.e. 24% by spade, 23% by sickles, 6% by bullock carts, 6% by manually operated fodder cutters, 5% each by power operated fodder cutters, diesel engines and tractors. Zhou & Roseman (1994) found both of the limbs as the most injured body part for 1000 farmers in Alabama by agricultural accidents. Mittal et al. (1996) have reported 36 agricultural injuries, of which 8.3% as fatal and 91.7% as non-fatal, in 12 villages of Punjab in India during 1 year. The injury rate per thousand machines per year was found to be 23.7 for tractors, 15.5 for sprayers, 7.1 for electric motors, 5.7 for threshers and 2.2 for fodder cutters. DeMuri and Purschwitz, (2000) have reported that the agricultural injuries are due to the lack of proper supervision, irrational expectations, economic difficulty and the lack of proper safety device. Huiyun Xiang et al. (2000) have considered 1500 Chinese farmers from 14 villages and conducted face to face interview of 1358 farmers between July 1997 and September 1997, to evaluate the agricultural related injuries. Rautiainen and Reynolds (2002) have carried out agricultural survey in the United States of America (USA) and they reported a high fatality rate of 0.22 per 1000 workers per year and the injuries rate as 5 to 166 per 1000 workers per year. A survey was conducted in selected villages of four states viz. Madhya Pradesh, Tamil Nadu, Orissa and Punjab on accidents happened during the year 1995-99 (Gite and Kot, 2003). The limited data collected indicated that the fatalities due to agricultural accidents were 21.2 per 100,000 workers per year. The major source of accidents were farm machines namely tractors, threshers, chaff cutters, cane crushers, sprayers and electric motors, and other sources namely snake bites, drowning in wells/ponds and lightening. Tiwari et al. (2002) have found the agricultural accident rate as 1.25 per thousand workers per year in Madhya Pradesh district in India. It is reported that 77.6% of agricultural accidents were because of farming machines, 11.8% because of hand tools and 10.6% due to others. Helkamp and Lundstrom (2002) have reported the injuries of farmers to be higher than industrial workers. In India, agricultural workers constitute as one of the important sources of farm power. Besides, they also operate animal drawn equipment, tractors, power tillers, self propelled and power operated machines. Table 1 gives the population dynamics of Indian agricultural workers which shows that by 2020 the population of agricultural workers in the country will be about 242 million of which 50% will be the female workers. Thus, there is going to be a significant role of farm workers in country's agriculture and due attention needs to be given to their safety and occupational health issues so as to have higher productivity, less accidents, and minimum occupational health problems.

Kumar et al. (2008) have found 576 agricultural related injuries with 332 tools such as 58% hand tool related, from 9 villages with 19,723 persons in the first phase. Further, in the second phase with 21 more villages of 78,890 persons, it was reported that 54 hand tools i.e.19% of hand tool was related out of 282 injuries. It was also recommended to have interventions development and training at block levels about the safety measures of equipments. Kumar et al. (2009) have investigated the agricultural accident for six years i.e. between years 2000-2005 of 42 villages of 4 districts in Arunachal Pradesh in India. It was reported to have the accident rate as 6.39 per thousand workers per year with 40% farm implement related injury.

Patel et al. (2010) have reported the agricultural accident rate as 0.8 per 1000 workers per year in Etawah district of Uttar Pradesh in India. Also it is reported the lack of study in agricultural injuries in developing countries due to non-availability of compiled information. Nilsson et al. (2010) have analyzed the responses from 223 injured farmers collected by the Swedish Farm Registry as part of a survey sent to 7,000 farms by the Swedish University of Agricultural Sciences and Statistics Sweden in 2004. These data showed that there were no significant differences in injuries incurred between the age groups, but that senior farmers seemed to suffer longer from their injuries. This study highlighted the importance of advising senior farmers to bear in mind that their bodies are no longer as young and strong as before. All age groups of farmers should, of course, be careful and consider the risks involved in their work, but since aging bodies need longer to heal, senior farmers probably need to be even more careful and review their work situation and work environment in order to avoid injuries during the final years of their working life. Then, It is recommended to educate senior farmers about the risks of injuries causing increasing damage due to their age. Copuroglu et al. (2012) have studied the agricultural injuries of 41 patients for a period of 3 years and found hand as the most commonly injured part compared to the lower limb and foot. Wibowo & Soni, (2016) have studied the farmers of East Java and Indonesia. Most of the farmer's injury was found in hand and the fatigue was reported as 92.8% in upper back, 93.6% in mid back and 91.8% in lower back. Safe, good and fit in hand were suggested as design criteria for hand tools. The recommended tool handles length was 12.4 cm and diameter was 3cm, respectively

RESEARCH METHODOLOGY

A standard questionnairee is designed and opinions are collected from experts of Agricultures, practioners and researchers and from literature review it is found that six most important risk factors of Indian Farming sector are

- Price or Market Risk
- Environmental &Human or Personal
- Legal / Policy Risk
- Resource Risk
- Health Risks
- Assets Risks
- Technology Risk

Then the Maut method is implemented to priortize the most important risk of the Agricultural sector in India. Then the risk factors are predicted with respect to out put Economic Growth of the Country and Environmental Conditions. Economic Growth and Environmental factor are most important factors for sustainability in Agricultural Business.

Table 1 shows selected factors for agricultural risk factor.

Table 2 shows correlation analysis.

Table 3 shows the item regression analysis of selected items.

Then Maut (Multi-Attribute Utility Theory) method is implemented for ranking dimensions of risk factors of three types of Agricultural sector (i.e.nominal Agrisector,Medium-sized Agrisector,Advanced Agrisector) methods.

Maut method implemented for ranking F the set of q criterion

Table 1. Selected factors for Agricultural Risk factor

1	Product recalls, defective products, rating agencies
2	Poor market timing, inadequate customer support
3	Natural hazards, facilities, disease outbreaks
4	Health, contract terms, turnover
5	No pollution and dust
6	No injury
7	War, terrorism, civil unrest, law, governing agencies
8	Reporting and compliance, environmental, food safety, traceability
9	Debt servicing, leverage, liquidity, solvency, profitability
10	Cash, interest rates, foreign exchange
11	information asymmetries, adverse selection
12	Cost, transportation, service availability, hold-up
13	Asset specificity, research and development
14	Complexity, obsolescence, workforce skill-sets, adoption rate, diffusion rate

Table 2. Correlation Analysis

	Price or Market Risk	Environmental &Human or Personal	Legal / Policy Risk	Resource Risk	Health Risks	Assets Risks &Technology Risk
Price or Market Risk	1					
Environmental &Human or Personal	0.351 0.003	1				
Legal / Policy Risk	0.343 0.004	0.400 0.001	1			
Resource Risk	-0.067 0.584	-0.033 0.791	-0.074 0.546	1		
Health Risks	-0.083 0.500	-0.333 0.005	-0.346 0.004	-0.159 0.192	1	
Assets Risks &Technology Risk	-0.313 0.009	-0.007 0.955	0.031 0.800	0.245 0.003	-0.116 0.344	1

Table 3. Regression Analysis

Item Analysis					
Omitted Variable	Adj.		Squared		
	Adj. Total Mean	Total StDev	Item-Adj. Corr	Total Corr	Multiple Cronbach's Alpha
C1	162.43	20.01	0.58432	1.00000	0.85875
C2	162.52	20.00	0.67233	1.00000	0.85787
C3	162.75	20.04	0.50814	1.00000	0.85993
C4	162.67	20.25	0.39919	1.00000	0.86230
C5	162.63	19.96	0.63080	1.00000	0.85793
C6	162.55	19.88	0.66241	1.00000	0.85702
C7	162.79	20.37	0.32343	1.00000	0.86365
C8	162.70	20.22	0.43469	1.00000	0.86171
C9	162.31	20.16	0.48088	1.00000	0.86086
C10	162.58	20.17	0.45174	1.00000	0.86127
C11	162.34	20.26	0.41167	1.00000	0.86218
C12	162.40	20.00	0.58540	1.00000	0.85870
C13	162.49	20.11	0.49500	1.00000	0.86042
C14	162.51	20.24	0.45072	1.00000	0.86164

$f_i (i = 1,2,3............q)$

Alternative $f_j (a_i)$ are first transferred into marginal utility contribution denoted by U_j, in order to avoid scale problems. The marginal utility scores are then aggregated with a weighted sum or addition (called additive module)

Utility Function can be written as -

$$\forall a_i \in A; U\left(a_i\right) = U\left[f_i\left(a_i\right),...f_q\left(a_i\right)\right]$$

$$= \sum_{j=1}^{q} U_j\left(f_i\left(a_i\right)\right).w_j$$

Where $U_j (f_j) \geq 0$ is usually a non-decreasing function and W_j represents the weight of criterion f_i

$$\sum_{j=1}^{q} w_i = 1$$

Marginal Utility functions are such that the best alternatives (Virtual or Real) on a specific criterion has a marginal utility score of an alternative is always 0 & 1.
Maximize the criterion

$$f_j{}'(a_i) = 1 + \left[\frac{f_j(a_i) - \min(f_j)}{\max(f_i) - \min(f_i)} \right] \tag{1}$$

Minimize the criterion

$$f_j{}'(a_i) = 1 + \left[\frac{\min(f_j) - f_j(a_i)}{\max(f_i) - \min(f_i)} \right] \tag{2}$$

Marginal Utility Score

$$U_i(a_i) = \frac{\exp\left(f_j(a_i^2) - 1\right)}{1.7} \tag{3}$$

Maut method is followed by calculating the Maximization and minimization criterion in table.4 by using equation 1 and 2. Table 5 shows the Marginal utility score by equation 3.Table.6 shows the score of nominal Agrisector, Medium-sized Agri sector, Advanced Agrisector.Then the closeness coefficient is calculated by using positive and negative scores and last Table.8 shows the ranks nominal Agrisector, Medium-sized Agri sector, Advanced Agrisector ranked Ist, it means it avoids maximum conditions of risk factors than other sectors. So this result will provide an insight into nominal Agrisector, Medium-sized sectors such that they will frame their policies, rules, and regulations to avoid risk factors in farming.Table.4 shows Maximizaion and Minimization criterion.

Table 5 shows RankedMarginal Utility Score.

Table 6 shows Score of Agrisectors.

Positive score Di+

0.9615 0.8967 0.8363 0.7005 0.7700 0.8120 D1+ 4.9771
0.9696 0.9126 0.8517 0.7234 0.7991 0.8135 D2+ 5.0698

Table 4. Maximizaion and Minimization criterion

	Price or Market Risk	Environmental & Human or Personal	Legal / Policy Risk	Resource Risk	Health Risks	Assets Risks & Technology Risk
Nominal Agrisector	0.0106 0.0193 0.0280	0.0124 0.0519 0.0914	0.0063 0.0846 0.1522	0.0256 0.1529 0.2803	0.0032 0.1175 0.2318	0.0216 0.0950 0.1684
Medium Agrisector	0.0076 0.0152 0.0228	0.0163 0.0360 0.0557	0.0249 0.0692 0.1522	0.0076 0.1301 0.2525	0.0057 0.0884 0.1711	0.0076 0.0935 0.1795
Advanced Agrisector	0.0106 0.0193 0.0280	0.0124 0.0519 0.0914	0.0121 0.0822 0.1522	0.0256 0.1529 0.2803	0.0032 0.1175 0.2318	0.0164 0.0924 0.1684

Table 5. RankedMarginal Utility Score

	Price or Market Risk	Environmental & Human or Personal	Legal / Policy Risk	Resource Risk	Health Risks	Assets Risks & Technology Risk
Nominal Agrisector	0.980759	0.948624	0.920891	0.853449	0.88746	0.906998
Medium Agrisector	0.984793	0.948624	0.920891	0.853449	0.88746	0.906998
Advanced Agrisector	0.980759	0.948624	0.920891	0.853449	0.88746	0.906998

Table 6. Score of Agrisectors.

	Price or Market Risk	Environmental & Human or Personal	Legal / Policy Risk	Resource Risk	Health Risks	Assets Risks & Technology Risk
Nominal Agrisector	0.020533	0.061133	0.100612	0.184913	0.150022	0.11232
Medium Agrisector	0.01644	0.039426	0.097622	0.16407	0.111241	0.116936
Advanced Agrisector	0.020533	0.061133	0.100126	0.184913	0.150022	0.111306

0.9615 0.8967 0.8387 0.7005 0.7700 0.8146 D3+ 4.9821

Negative score Di-

0.0013 0.0092 0.0161 0.0320 0.0326 0.0173 D1- 0.1084
0.0012 0.0034 0.0284 0.0340 0.0228 0.0234 D2- 0.1133
0.0013 0.0092 0.0180 0.0320 0.0326 0.0189 D3- 0.1119

Table 7 shows Closeness Coefficient.

Among all risk factors of Agricultural business resource risk ranked ist, Assets Risks &Technology Risk ranked first resource risk 3rd and health risk second for Indian farmers. Assets risk and technology risk is the risk factors can not be solved or resolved by an individual farmer or farmer community, Agripolicies and Govt. attention towards agribusiness can resolve this risk in terms of new and advanced equipment, proper supply chain distribution, by marketing policies and agricultural policies towards loan and subsidies provided to farmers. Resource risk also include the seeds, pesticides, lands, water provided for irrigation. Health risk can be resolved by providing safety training and security to farmers.

CONCLUSION

Risk management has gained relevance in agriculture due to growing risks (for instance, agricultural and input price volatility, climate change), the limited and often decreasing risk-bearing capacity of farms and the intention of the majority of farmers to limit their farms' exposure to risks. Therefore, a systematic risk management process should be implemented and regularly performed on future-oriented farms. The importance of systematic risk management grows the more non-family workers are hired, the higher the debt to equity ratio, and the higher the share of lease land. The Business of agriculture is subject to big and many suspicions and uncertainties. So far maximum people in India make their source of revenue from this agribusiness compared to from all other sectors put together. Understanding

Table 7. Closeness Coefficient

Closeness Coeficient		Type	Rank
Cc1	0.021322	Nominal Agrisector	3
Cc2	0.02186	Medium Agrisector	2
Cc3	0.021968	Advanced Agrisector	1

agricultural risk factors and the ways of managing it is, therefore, the most important topic of discussion that deserves serious concentration and investigates. Despite its noticeable significance, risk factor management in agriculture is an under-researched topic relative to traditional concerns.

REFERENCES

Agarwal, A., Shankar, R., & Tiwari, M. K. (2006). Modeling agility of supply chain. *Industrial Marketing Management, 36*, 443–457.

Akao, Y. (1990). *Quality function deployment: Integrating customer requirements into product design*. Cambridge, MA: Journal of Productivity Press.

Akao, Y. (1997). QFD: Past, Present, and Future. *Proceedings of the International Symposium on QFD*.

Banthia, J. K. (2004). *Census of India 2001 - Primary Census Abstracts*. Registrar General & Census Commissioner, Govt. of India.

Chan, L. K., & Wu, M. L. (2002). Quality function deployment: A literature review. *European Journal of Operational Research, 143*(3), 463–497.

Chan, L. K., & Wu, M. L. (2005). A systematic approach to quality function deployment with a full illustrative example. *International Journal of Management Sciences, 33*, 119–139.

Chen, L. H., & Weng, M. C. (2006). An evaluation approach to engineering design in QFD processes using fuzzy goal programming models. *European Journal of Operational Research, 172*, 230–248.

Cohen, L. (1995). *Quality function deployment: How to make QFD work for You*. Reading, MA: Addison- Wesley.

Cooper, S. P., Burau, K. E., & Frankowski, R. (2006). A cohort study of injuries in migrant farm worker families in South Texas. *Annals of Epidemiology, 16*(4), 313–320.

Copuroglu, C., Heybeli, N., Ozcan, M., Yilmaz, B., Ciftdemir, M., and Copuroglu, E. (2012). Major extremity injuries associated with farmyards accidents. *The Scientific World Journal.* doi:10.1100/2012/314038

Cristiano, J. J., Liker, J. K., & White, Ch. C. (2000). Customer-driven product development through quality function deployment in the U.S. and Japan. *Journal of Product Innovation Management, 17*(4), 286–308.

DeMuri, G. P., & Purschwitz, M. A. (2000). Farm injuries in children: A review. *Wisconsin Medical Journal, 99*(9), 51–55.

Erdogan, B. (2003). *The extent of information visualization in Turkish construction industry, A QFD approach.* The Middle East Technical University.

Fisher, C., & Schutta, J. T. (2003). *Developing New Service- Incorporating the Voice of the Customer into Strategic Service Development.* Milwaukee, WI: ASQ Quality Press.

Griffin, A., & Hauser, J. R. (1992). Patterns of communications among marketing, engineering and manufacturing. *Journal of Management and Science, 38*(3), 360–373.

Griffin, A., & Hauser, R. J. (1991). The Voice of the Customer. *Marketing Science, 12*(1), 1–27.

Guinta, L. R., & Praizler, N. C. (1993). *The QFD Book, The team approaches to solving problems and satisfying customers through quality function deployment.* New York: American Management Association.

Hauser, J., & Clausing, D. (1989). The house of quality. *Harvard Business Review, 66*(3), 63–73.

Helkamp, J., & Lundstrom, W. (2002). Tractor Related Deaths Among West Virginia Farmers. *Annals of Epidemiology, 12*(7), 510. doi: 10.1016/S1047-2797(02)00344-7

Hope, A., Kelleher, C., Holmes, L., & Hennessy, T. (1999). Health and safety practices among farmers and other workers: A needs assessment. *Occupational Medicine (Philadelphia, Pa.), 49,* 231–235.

Huiyun, X., Zengzhen, W., Lorann, S., Thomas, J. K., Xuzhen, H., & Xianghua, F. (2000). Article. *American Journal of Public Health, 90,* 1269–1276.

Karthikeyan, C., Veeraragavathatham, D., Karpagam, D., & Ayisha, F. S. (2009). Traditional tools in agricultural practices. *Indian Journal of Traditional Knowledge, 8*(2), 212–217.

Knapp, L. W. (1965). Agricultural Injuries Prevention. *Journal of Occupational Medicine., 7*(11), 553–745.

Knapp, L. W. (1966). Occupational and Rural Accidents. *Archives of Environmental Health, 13,* 501–506.

Kogure, M., & Akao, Y. (1983). Quality function deployment and CWQC in Japan. *Journal of Quality and Progress, 16*(10), 25–29.

Kumar, A., Singh, J. K., Mohan, D., & Varghese, M. (2008). Farm hand tools injuries: A case study from Northern India. *Safety Science, 46*, 54–65. doi: 10.1016/j. ssci.2007.03.003

Kumar, A., Varghese, M., & Mohan, D. (2000). Equipment-related injuries in agriculture: An international perspective. *Injury Control and Safety Promotion, 7*(3), 1–12.

Lombardi, M., & Fargnoli, M. (2018). Article. *International Journal of Safety and Security Engineering, 8*(2), 342–353.

Maritan, D., & Panizzolo, R. (2009). Identifying business priorities through quality function deployment. *Marketing Intelligence & Planning, 27*(5), 714–728.

Mishra, D., & Satapathy, S. (2018). Drudgery Reduction of Farm Women of Odisha by Improved Serrated Sickle. *International Journal of Mechanical Engineering and Technology, 9*(2), 53–61.

Mittal, V. K., Bhatia, B. S., & Ahuja, S. S. (1996). *A Study of the Magnitude, Causes, and Profile of Victims of Injuries with Selected Farm Machines in Punjab. Final Report of ICAR adhoc Research Project.* Ludhiana, India: Department of Farm Machinery and Power Engineering, Punjab Agricultural University.

Mohan, D., & Patel, R. (1992). Design of safer agricultural equipment: Application of ergonomics and epidemiology. *International Journal of Industrial Ergonomics, 10*, 301–309. doi: 10.1016/0169-8141(92)90097-J

Murphy, D. J. (1992). *Safety and Health for Production Agriculture.* St. Joseph, MI: American Society of Agricultural Engineers.

Nag, P. K., & Nag, A. (2004). Drudgery, accidents and injuries in Indian agriculture. *Industrial Health, 42*(4), 149–162. doi: 10.2486/indhealth.42.149

Patel, S. K. (2010). Agricultural injuries in Etawah district of Uttar Pradesh in India. *Safety Science, 48*, 222–229. https://doi.org/10.1016/j.ssci.2009.08.003

Pawitra, T. A., & Tan, K. C. (2003). Tourist satisfaction in Singapore- a perspective from Indonesian tourists. *Managing Service Quality, 13*(5), 339–411.

Pradhan, C. K., Goswami, A., Ghosh, S. N., & Nag, P. K. (1986). Evaluation of working with spade in agriculture. *The Indian Journal of Medical Research, 84*, 424–429.

Prasanna Kumar, G. V., & Dewangan, K. N. (2009). Agricultural accidents in north eastern region of India. *Safety Science*, *47*, 199–205. https://doi.org/10.1016/j.ssci.2008.03.007

Raj, T., & Attri, R. (2011). Identification and modeling of barriers in the implementation of TQM. *International Journal of Productivity and Quality Management.*, *28*(2), 153–179. doi: 10.1504/IJPQM.2011.041844

Rautiainen, R. H., & Reynolds, S. J. (2002). Mortality and morbidity in agriculture in the United States. *Journal of Agricultural Safety and Health*, *8*(3), 259–276. doi:10.13031/2013.9054

Ravi, V., & Shankar, R. (2005). Analysis of interactions among the barriers of reverse logistics. *Technological Forecasting and Social Change*, *72*, 1011–1029. doi: 10.1016/j.techfore.2004.07.002

Robertoes Koekoeh, K. W., & Peeyush, S. (2016). Agriculture and Agricultural Science. *Procedia*, *9*, 323 – 327. doi: 10.1016/j.aaspro.2016.02.142

Satapathy, S. (2014). ANN, QFD and ISM approach for framing electricity utility service in India for consumer satisfaction. *Int. J. Services and Operations Management*, *18*(4), 404–428. doi: 10.1504/IJSOM.2014.063243

Tan, K. C., & Pawitra, T. A. (2001). Integrating SERVQUAL and Kana's model into QFD for service excellence development. *Managing Service Quality*, *11*(6), 418–430.

Voaklander, K. D., Kelly, D. C., & Rowe, B. H. (2006). Pain, medication, and injury in older farmers. *American Journal of Industrial Medicine*, *49*(5), 374–382. doi: 10.1002/ajim.20292

Zhou, C., & Roseman, J. M. (1994). Agricultural injuries among a population-based sample of farm operators in Alabama. *American Journal of Industrial Medicine*, *25*(3), 385–402. doi: 10.1002/ajim.4700250307

KEY TERMS AND DEFINITIONS

ARM Agricultural Risk Management: AARM is an innovative approach for improving the resilience of vulnerable rural households and leveraging finance and investment. ARM allows farmers and businesses to be pro-active and increases their capacity to assess, prepare for, absorb, and adapt to risks.

Fuzzy Rule: Fuzzy rule is a conditional statement. The form of fuzzy rules is given by IF THEN statements. If y is B THEN x is A, where x and y are linguistic variables, A and B are linguistic values determined by fuzzy sets.

Fuzzy Set: Fuzzy set is expressed as a function and the elements of the set are mapped into their degree of membership. A set with the fuzzy boundaries are 'hot', 'medium', or 'cold' for temperature.

Fuzzy Variable: A quantity that can take on linguistic values. For example, the fuzzy variable 'disease' might have values such as 'low', 'medium', or 'high'.

MAUT Method: MAUT method is implemented for ranking dimensions. Multi-attribute utility theory (MAUT) is a structured methodology designed to handle the tradeoffs among multiple objectives. One of the first applications of MAUT involved a study of alternative locations for a new airport in Mexico City in the early 1970s. The factors that were considered included cost, capacity, access time to the airport, safety, social disruption, and noise pollution.

Platform for Agricultural Risk Management (PARM): PARM has the global mandate to contribute to sustainable agricultural growth: boost rural investment, reduce food insecurity, and improve resilience to climate change.

Compilation of References

\Dobkin & Williams. (2012). *Analog Circuit Design Volume 2: Immersion in the Black Art of Analog Design*. Newnes.

Abayomi, A. A., Ikuomola, A. J., Aliyu, O.A., & Alli, O. A., (2014). Development of a Mobile Remote Health Monitoring system – MRHMS. *IEEE African Journal of Computing & ICT*, *7*(4), 14-22,

Abernethy, R. B. (2006). The New Weibull Handbook (5th ed.). Academic Press.

Agarwal, S., & Yadav, R. S. (2008). System level energy aware fault tolerance approach for real time systems. *IEEE Region 10 Conference, TENCON*, 1-6.

Agarwal, A., Shankar, R., & Tiwari, M. K. (2006). Modeling agility of supply chain. *Industrial Marketing Management*, *36*(4), 443–457. doi:10.1016/j.indmarman.2005.12.004

Agarwalla, P., & Mukhopadhyay, S. (2020). Hybrid advanced player selection strategy based population search for global optimization. *Expert Systems with Applications*, *139*, 112825. doi:10.1016/j.eswa.2019.112825

Aggarwal, A. G., Kapur, P., Kaur, G., & Kumar, R. (2012). Genetic algorithm based optimal testing effort allocation problem for modular software. *Bharati Vidyapeeth's Institute of Computer Applications and Management (BVICAM)*, 445.

Aggarwal, A. G., Dhaka, V., & Nijhawan, N. (2017). Reliability analysis for multi-release open-source software systems with change point and exponentiated Weibull fault reduction factor. *Life Cycle Reliability and Safety Engineering*, *6*(1), 3–14. doi:10.100741872-017-0001-0

Aggarwal, A. G., Gandhi, N., Verma, V., & Tandon, A. (2019). Multi-Release Software Reliability Growth Assessment: An approach incorporating Fault Reduction Factor and Imperfect Debugging. *International Journal of Mathematics in Operational Research.*, *15*(4), 446. doi:10.1504/IJMOR.2019.103006

Ahmadpour, A., & Torkzadeh, P. (2012). An Enhanced Bulk-Driven Folded-Cascode Amplifier in 0.18 μm CMOS Technology. *Circuits and Systems*, 187-191.

Ahmadreza, R. (1996). A 1-GHz CMOS RF front-end IC for a direct-conversion wireless receiver. *IEEE Journal of Solid-State Circuits*, *31*(7), 880–889. doi:10.1109/4.508199

Ahmad, S. W., & Bamnote, G. (2019). Whale–crow optimization (WCO)-based Optimal Regression model for Software Cost Estimation. *Sadhana*, *44*(4), 94. doi:10.100712046-019-1085-1

Akao, Y. (1997). QFD: Past, Present, and Future. *Proceedings of the International Symposium on QFD.*

Akao, Y. (1990). *Quality function deployment: Integrating customer requirements into product design.* Cambridge, MA: Journal of Productivity Press.

Akturk, I., & Karpuzcu, U. R. (2017). AMNESIAC: Amnesic Automatic Computer. *Proceedings of the Twenty-Second International Conference on Architectural Support for Programming Languages and Operating Systems*, 811-824.

Al-Betar, M. A., Awadallah, M. A., Faris, H., Aljarah, I., & Hammouri, A. I. (2018). Natural selection methods for grey wolf optimizer. *Expert Systems with Applications*, *113*, 481–498. doi:10.1016/j.eswa.2018.07.022

Aljahdali, S., & Sheta, A. F. (2011). *Predicting the reliability of software systems using fuzzy logic.* Paper presented at the 2011 Eighth International Conference on Information Technology: New Generations. 10.1109/ITNG.2011.14

Al-Rahamneh, Z., Reyalat, M., Sheta, A. F., Bani-Ahmad, S., & Al-Oqeili, S. (2011). A new software reliability growth model: Genetic-programming-based approach. *Journal of Software Engineering and Applications*, *4*(8), 476–481. doi:10.4236/jsea.2011.48054

AL-Saati, D., Akram, N., & Abd-AlKareem, M. (2013). *The use of cuckoo search in estimating the parameters of software reliability growth models.* arXiv preprint arXiv:1307.6023

Anand, S., Verma, V., & Aggarwal, A. G. (2018). 2-Dimensional Multi-Release Software Reliability Modelling considering Fault Reduction Factor under imperfect debugging. *INGENIERIA SOLIDARIA, 14.*

Anisheh, S. M., & Shamsi, H. (2017). Two-stage class-AB OTA with enhanced DC gain and slew rate. *International Journal of Electronics Letters*, 4.

Anonymous. (2014). *MSDN Windows Presentation Foundation.* Retrieved from http://msdn. microsoft.com/en-us/library/vstudio/ms754130.aspx

Ascher, H., & Feingold, H. (1978). *Application of Laplace's test to repairable system reliability.* Paper presented at the Actes du 1er Colloque International de Fiabilité et de Maintenabilité.

Ascher, H., & Feingold, H. (1984). *Repairable systems reliability: modeling, inference, misconceptions and their causes.* M. Dekker.

Ayav, T., Fradet, P., & Girault, A. (2006). Implementing Fault Tolerance in Real-Time Systems by Automatic Program Transformations. *Proceedings of the 6th ACM & IEEE International conference on Embedded software*, 205-214. 10.1145/1176887.1176917

Azizi, R. (2014). *Empirical study of artificial fish swarm algorithm.* arXiv preprint arXiv:1405.4138

Banati, H., & Chaudhary, R. (2017). Multi-modal bat algorithm with improved search (MMBAIS). *Journal of Computational Science, 23*, 130–144. doi:10.1016/j.jocs.2016.12.003

Bansal, J. C., Sharma, H., Jadon, S. S., & Clerc, M. (2014). Spider monkey optimization algorithm for numerical optimization. *Memetic Computing, 6*(1), 31-47.

Banthia, J. K. (2004). *Census of India 2001 - Primary Census Abstracts.* Registrar General & Census Commissioner, Govt. of India.

Barger, P., Schön, W., & Bouali, M. (2009). *A study of railway ERTMS safety with Colored Petri Nets.* The European Safety and Reliability Conference 2009, Prague, Czech Republic.

Basu, M., & Chowdhury, A. (2013). Cuckoo search algorithm for economic dispatch. *Energy, 60*, 99–108. doi:10.1016/j.energy.2013.07.011

Baunmann, R. (2005). Soft errors in advanced computer systems. *IEEE Transactions on Design & Test of Computers, 22*(3), 258-266.

Baysari, M. T., Caponecchia, C., McIntosh, A. S., & Wilson, J. R. (2009). Classification of errors contributing to rail incidents and accidents: A comparison of two human error identification techniques. *Safety Science, 47*(7), 948–957. doi:10.1016/j.ssci.2008.09.012

Bertsche, B. (2010). *Reliability in Automotive and Mechanical Engineering: Determination of Component and System Reliability.* Berlin: Springer.

Bonabeau, E., Marco, D. D. R. D. F., Dorigo, M., & Theraulaz, G. (1999). *Swarm intelligence: from natural to artificial systems (No. 1).* Oxford University Press.

Bracchi, P., & Cukic, B. (2004). Performability modeling of mobile software systems. In *Proceedings of the International Symposium on Software Reliability Engineering* (pp. 77-84). Bretange, France: IEEE. 10.1109/ISSRE.2004.27

Briggs, P., Cooper, K. D., & Torczon, L. (1992). Rematerialization. *Proceedings of the 92nd Programming Language Design and Implementation.*

Cabrera-Bernal, E., Pennisi, S., Grasso, A. D., Torralba, A., & Carvajal, R. G. (2016). 0.7-V Three-Stage Class-AB CMOS Operational Transconductance Amplifier. *IEEE Transactions on Circuits and Systems, 36*, 1807-1815.

Cacciabue, P. C. (2005). Human error risk management methodology for safety audit of a large railway organisation. *Applied Ergonomics, 36*(6), 709–718. doi:10.1016/j.apergo.2005.04.005 PMID:16122693

Callet, S., Fassi, S., Fedeler, H., Ledoux, D., & Navarro, T. (2014). *The use of a "model-based design" approach on an ERTMS Level 2 ground system. In Formal Methods Applied to Industrial Complex Systems* (pp. 165–190). Hoboken, NJ: John Wiley & Sons.

Castillo, E., Calviño, A., Grande, Z., Sánchez-Cambronero, S., Rivas, I. G. A., & Menéndez, J. M. (2016). A Markovian–Bayesian network for risk analysis of high speed and conventional railway lines integrating human errors. *Computer-Aided Civil and Infrastructure Engineering*, *31*(3), 193–218. doi:10.1111/mice.12153

Castillo, E., Grande, Z., Mora, E., Lo, H. K., & Xu, X. D. (2017). Complexity reduction and sensitivity analysis in road probabilistic safety assessment Bayesian network models. *Computer-Aided Civil and Infrastructure Engineering*, *32*(7), 546–561. doi:10.1111/mice.12273

Celik, Y., & Ulker, E. (2013). An improved marriage in honey bees optimization algorithm for single objective unconstrained optimization. *The Scientific World Journal*. PMID:23935416

Chabridon, S., & Gelenbe, E. (1995). Failure detection algorithms for a reliable execution of parallel programs. *Proc. International Symposium on Reliable Distributed Systems*, 229-238. 10.1109/RELDIS.1995.526230

Chanda, U., Tandon, A., & Kapur, P. K. (2010). Software Release Policy based on Change Point Considering Risk Cost. *Advances in Information Theory and Operations Research*, 111-122.

Chang, H. W. D., & Oldham, W. J. B. (1995). Dynamic task allocation models for large distributed computing systems. *IEEE Transactions on Parallel and Distributed Systems*, *6*(12), 1301–1315. doi:10.1109/71.476170

Chang, I. H., Pham, H., Lee, S. W., & Song, K. Y. (2014). A testing-coverage software reliability model with the uncertainty of operating environments. *International Journal of Systems Science: Operations & Logistics*, *1*(4), 220–227.

Chan, L. K., & Wu, M. L. (2002). Quality function deployment: A literature review. *European Journal of Operational Research*, *143*(3), 463–497.

Chan, L. K., & Wu, M. L. (2005). A systematic approach to quality function deployment with a full illustrative example. *International Journal of Management Sciences*, *33*, 119–139.

Chatterjee, S., & Maji, B. (2018). A bayesian belief network based model for predicting software faults in early phase of software development process. *Applied Intelligence*, 1–15.

Chatterjee, S., & Shukla, A. (2016). Modeling and analysis of software fault detection and correction process through weibull-type fault reduction factor, change point and imperfect debugging. *Arabian Journal for Science and Engineering*, *41*(12), 5009–5025. doi:10.100713369-016-2189 0

Chatterjee, S., & Shukla, A. (2017). An ideal software release policy for an improved software reliability growth model incorporating imperfect debugging with fault removal efficiency and change point. *Asia-Pacific Journal of Operational Research*, *34*(03), 1740017. doi:10.1142/S0217595917400176

Chaudhary, A., Agarwal, A. P., Rana, A., & Kumar, V. (2019). *Crow Search Optimization Based Approach for Parameter Estimation of SRGMs*. Paper presented at the 2019 Amity International Conference on Artificial Intelligence (AICAI). 10.1109/AICAI.2019.8701318

Chean, M., & Fortes, J. (1990). The full-use-of-suitable-spares (FUSS) approach to hardware reconfiguration for fault-tolerant processor arrays. *IEEE Transactions on Computers*, *39*(4), 564–571. doi:10.1109/12.54851

Cheema, A., & Singh, M. (2019, April). An application of phonocardiography signals for psychological stress detection using non-linear entropy-based features in empirical mode decomposition domain. *Applied Soft Computing*, *77*, 24–33. doi:10.1016/j.asoc.2019.01.006

Chen, G., Kandemir, M., & Li, F. (2006). Energy-Aware Computation Duplication for Improving Reliability in Embedded Chip Mulitprocessors. *Asia and South Pacific Conference on Design Automation*, 134-139.

Chen, L. H., & Weng, M. C. (2006). An evaluation approach to engineering design in QFD processes using fuzzy goal programming models. *European Journal of Operational Research*, *172*, 230–248.

Chen, Y., Li, L., Peng, H., Xiao, J., Yang, Y., & Shi, Y. (2017). Particle swarm optimizer with two differential mutation. *Applied Soft Computing*, *61*, 314–330. doi:10.1016/j.asoc.2017.07.020

Chevochot, P., & Puaut, I. (1999). Scheduling Fault-Tolerant Distributed Hard-Real Time Tasks Independently on the Replication Strategies. *6th Int. Conf. on Real-Time Comp. Syst. And Applications*, 356-363. 10.1109/RTCSA.1999.811280

Choudhary, A., Baghel, A. S., & Sangwan, O. P. (2017). An efficient parameter estimation of software reliability growth models using gravitational search algorithm. *International Journal of System Assurance Engineering and Management*, *8*(1), 79–88. doi:10.100713198-016-0541-0

Choudhary, A., Baghel, A. S., & Sangwan, O. P. (2018). *Parameter Estimation of Software Reliability Model Using Firefly Optimization. In Data Engineering and Intelligent Computing* (pp. 407–415). Springer.

Ciardo, G., Muppala, J., & Trivedi, K. S. (1992). Analyzing concurrent and fault-tolerant software using stochastic reward net. *Journal of Parallel and Distributed Computing*, *15*(3), 255–269. doi:10.1016/0743-7315(92)90007-A

Cohen, L. (1995). *Quality function deployment: How to make QFD work for You*. Reading, MA: Addison- Wesley.

Cooper, S. P., Burau, K. E., Frankowski, R., Shipp, E., Deljunco, D., Whitworth, R., ... Hanis, C. (2006). A cohort study of injuries in migrant farm worker families in South Texas. *Annals of Epidemiology*, *16*(4), 313–320. doi:10.1016/j.annepidem.2005.04.004 PMID:15994097

Copuroglu, C., Heybeli, N., Ozcan, M., Yilmaz, B., Ciftdemir, M., and Copuroglu, E. (2012). Major extremity injuries associated with farmyards accidents. *The Scientific World Journal*. doi:10.1100/2012/314038

Cristiano, J. J., Liker, J. K., & White, Ch. C. (2000). Customer-driven product development through quality function deployment in the U.S. and Japan. *Journal of Product Innovation Management*, *17*(4), 286–308.

Csorba, M., Heegaard, P., & Hermann, P. (2008). Cost-Efficient deployment of collaborating components. In *Proceedings of the 8th IFIP WG 6.1 International conference on Distributed applications and interoperable systems* (pp. 253-268). Oslo, Norway: Springer. 10.1007/978-3-540-68642-2_20

Dave, B. P., & Jha, N. K. (1999). COFTA: Hardware-Software Co-Synthesis of Heterogeneous Distributed Embedded System for Low Overhead Fault Tolerance. *IEEE Transactions on Computers, 48*(4), 417–441. doi:10.1109/12.762534

DeMuri, G. P., & Purschwitz, M. A. (2000). Farm injuries in children: A review. *Wisconsin Medical Journal, 99*(9), 51–55.

DeMuri, G. P., & Purschwitz, M. A. (2000). Fatal injuries in children: A review. *WMJ: Official Publication of the State Medical Society of Wisconsin, 99*(9), 51–55. PMID:11220197

Deri, G., & Varro. (2008). Model Driven Performability Analysis of Service Configurations with Reliable Messaging. In *Proceedings of the Model-Driven Web Engineering*, (pp. 61-75). Toulouse, France: CEUR.

Dilmac, S., & Korurek, M. (2015). ECG heart beat classification method based on modified ABC algorithm. *Applied Soft Computing, 36*, 641–655. doi:10.1016/j.asoc.2015.07.010

Dohi, T. (2000). The age-dependent optimal warranty policy and its application to software maintenance contract. *Proceedings of 5th International Conference on Probabilistic Safety Assessment and Management.*

Dorigo, M., & Di Caro, G. (1999). Ant colony optimization: a new meta-heuristic. *Proceedings of the 1999 congress on evolutionary computation-CEC99.* 10.1109/CEC.1999.782657

Dorigo, M., Birattari, M., & Stutzle, T. (2006). Ant colony optimization. *Comput Intell Magaz, IEEE, 1*(4), 28–39. doi:10.1109/MCI.2006.329691

Dorrian, J., Roach, G. D., Fletcher, A., & Dawson, D. (2007). Simulated train driving: Fatigue, self-awareness and cognitive disengagement. *Applied Ergonomics, 38*(2), 155–166. doi:10.1016/j.apergo.2006.03.006 PMID:16854365

Eberhart, R., & Kennedy, J. (1995). A new optimizer using particle swarm theory. *Proceedings of the Sixth International Symposium on Micro Machine and Human Science.* 10.1109/MHS.1995.494215

Eberhart, R., & Kennedy, J. (1995, November). Particle swarm optimization. In *Proceedings of the IEEE international conference on neural networks* (Vol. 4, pp. 1942-1948). IEEE. 10.1109/ICNN.1995.488968

Efe, K. (1982). Heuristic models of task assignment scheduling in distributed systems. *Computer.*

Eles, P., Izosimov, V., Pop, P., & Peng, Z. (2008). Synthesis of Fault-Tolerant Embedded Systems. *Design, Automation and Test in Europe*, 1117-1122.

Emary, E., Zawbaa, H. M., & Hassanien, A. E. (2016). Binary grey wolf optimization approaches for feature selection. *Neurocomputing, 172*, 371–381. doi:10.1016/j.neucom.2015.06.083

Erdogan, B. (2003). *The extent of information visualization in Turkish construction industry, A QFD approach.* The Middle East Technical University.

Felipe, A., Sallak, M., Schön, W., & Belmonte, F. (2013). Application of evidential networks in quantitative analysis of railway accidents. *Proceedings of the Institution of Mechanical Engineers, Part O: Journal of Risk and Reliability, 227*(4), 368–384.

Fenton, N., Neil, M., & Marquez, D. (2008). Using Bayesian networks to predict software defects and reliability. *Proceedings of the Institution of Mechanical Engineers, Part O: Journal of Risk and Reliability, 222*(4), 701–712.

Fisher, C., & Schutta, J. T. (2003). *Developing New Service- Incorporating the Voice of the Customer into Strategic Service Development.* Milwaukee, WI: ASQ Quality Press.

Flammini, F., Marrone, S., & Mazzocca, N. (2006). *Modelling system reliability aspects of ERTMS/ ETCS by fault trees and Bayesian networks.* The European Safety and Reliability Conference 2006, Estoril, Portugal.

Flammini, F., Marrone, S., Iacono, M., Mazzocca, N., & Vittorini, V. (2014). A Multiformalism modular approach to ERTMS/ETCS failure modeling. *International Journal of Reliability Quality and Safety Engineering, 21*(1), 1450001. doi:10.1142/S0218539314500016

Flandre, S. (1996). A gm/ID based methodology for the design of CMOS analog circuits and its application to the synthesis of a silicon-on-insulator micropower OTA. *IEEE Journal of Solid-State Circuits, 31*(9), 1314–1319. doi:10.1109/4.535416

France, M. E. (2017) *Engineering for Humans: A New Extension to STPA, in Aeronautics and Astronautics* (Master thesis). Massachusetts Institute of Technology.

Friedman, M. A., Tran, P. Y., & Goddard, P. I. (1995). *Reliability of software intensive systems.* William Andrew.

Galan, J. A., López-Martín, A. J., Carvajal, R. G., Ramírez-Angulo, J., & Rubia-Marcos, C. (2007). Super Class-AB OTAs With Adaptive Biasing and Dynamic Output Current Scaling. *IEEE Transactions on Circuits and Systems.*

Gandomi, A. H., & Yang, X.-S. (2014). Chaotic bat algorithm. *Journal of Computational Science, 5*(2), 224–232. doi:10.1016/j.jocs.2013.10.002

Garg, H. (2015). An efficient biogeography based optimization algorithm for solving reliability optimization problems. *Swarm and Evolutionary Computation, 24*, 1–10. doi:10.1016/j.swevo.2015.05.001

Garg, H. (2016). A hybrid PSO-GA algorithm for constrained optimization problems. *Applied Mathematics and Computation, 274*, 292–305. doi:10.1016/j.amc.2015.11.001

Garg, H. (2017). Performance analysis of an industrial system using soft computing based hybridized technique. *Journal of the Brazilian Society of Mechanical Sciences and Engineering, 39*(4), 1441–1451. doi:10.100740430-016-0552-4

Garg, H. (2019). A hybrid GSA-GA algorithm for constrained optimization problems. *Information Sciences, 478*, 499–523. doi:10.1016/j.ins.2018.11.041

Garg, H., & Sharma, S. (2013). Multi-objective reliability-redundancy allocation problem using particle swarm optimization. *Computers & Industrial Engineering, 64*(1), 247–255. doi:10.1016/j.cie.2012.09.015

Gautam, S., Kumar, D., & Patnaik, L. (2018). *Selection of Optimal Method of Software Release Time Incorporating Imperfect Debugging.* Paper presented at the 4th International Conference on Computational Intelligence & Communication Technology (CICT). 10.1109/CIACT.2018.8480133

Ghazel, M. (2014). Formalizing a subset of ERTMS/ETCS specifications for verification purposes. *Transportation Research Part C: Engineering Technologies, 42*, 60–75. doi:10.1016/j.trc.2014.02.002

Girualt, A., Kalla, H., Sighireanu, M., & Sorel, Y. (2003). An Algorithm for Automatically Obtaining Distributed and Fault-Tolerant Static Schedules. *Int. Conf. on Dependable Syst. And Netw.*, 159-168. 10.1109/DSN.2003.1209927

Gite, L. P., & Kot, L. S. (2003). *Accidents in Indian Agriculture.* Technical Bulletin No. CIAE/2003/103. Coordinating Cell, All India Coordinated Research Project on Human Engineering and Safety in Agriculture, Central Institute of Agricultural Engineering, Nabibagh, Bhopal.

Glaß, M. Lukasiewycz, M., Streichert, T., Haubelt, C., & Teich, J., (2007). Reliability Aware System Synthesis. *Design, Automation & Test in Europe Conference & Exhibition*, 1-6.

Goel, A. L. (1985). Software reliability models: Assumptions, limitations, and applicability. *IEEE Transactions on Software Engineering, SE-11*(12), 1411–1423. doi:10.1109/TSE.1985.232177

Goel, A. L., & Okumoto, K. (1979). Time-dependent error-detection rate model for software reliability and other performance measures. *IEEE Transactions on Reliability, 28*(3), 206–211. doi:10.1109/TR.1979.5220566

GOI. (2002). *India Vision 2020. Planning commission.* New Delhi: Govt. of India.

GOI. (2006). *Population projections for India and States 2001-2006, Report of the technical group on population projections constituted by the National Commission on Population.* New Delhi: Govt. of India.

Gond, C., Melhem, R., & Gupta, R. (1996). Loop transformations for fault detection in regular loops on massively parallel systems. *IEEE Transactions on Parallel and Distributed Systems, 7*(12), 1238–1249. doi:10.1109/71.553273

Griffin, A., & Hauser, J. R. (1992). Patterns of communications among marketing, engineering and manufacturing. *Journal of Management and Science, 38*(3), 360–373.

Griffin, A., & Hauser, R. J. (1991). The Voice of the Customer. *Marketing Science*, *12*(1), 1–27.

Guest, G., Bunce, A., & Johnson, L. (2006). How many interviews are enough?: An experiment with data saturation and variability. *Field Methods*, *18*(1), 59–82. doi:10.1177/1525822X05279903

Guinta, L. R., & Praizler, N. C. (1993). *The QFD Book, The team approaches to solving problems and satisfying customers through quality function deployment*. New York: American Management Association.

Gupta, V., Singh, A., Sharma, K., & Mittal, H. (2018). *A novel differential evolution test case optimisation (detco) technique for branch coverage fault detection. In Smart Computing and Informatics* (pp. 245–254). Springer.

Hamilton, W. I., & Clarke, T. (2005). Driver performance modelling and its practical application to railway safety. *Applied Ergonomics*, *36*(6), 661–670. doi:10.1016/j.apergo.2005.07.005 PMID:16182232

Hasançebi, O., & Kazemzadeh Azad, S. (2012). An Efficient Metaheuristic Algorithm for Engineering Optimization: SOPT. *International Journal of Optimization in Civil Engineering*, 479-487.

Hassan, K., & Shamsi, H. (2010). A sub-1V high-gain two-stage OTA using bulk-driven and positive feedback techniques. *European Conference on Circuits and Systems for Communications*.

Hauser, J., & Clausing, D. (1989). The house of quality. *Harvard Business Review*, *66*(3), 63–73.

Heidari, A. A., & Pahlavani, P. (2017). An efficient modified grey wolf optimizer with Lévy flight for optimization tasks. *Applied Soft Computing*, *60*, 115–134. doi:10.1016/j.asoc.2017.06.044

Helkamp, J., & Lundstrom, W. (2002). Tractor Related Deaths Among West Virginia Farmers. *Annals of Epidemiology*, *12*(7), 510. doi:10.1016/S1047-2797(02)00344-7

Herrmann, P., & Krumm, H. (2000). A framework for modeling transfer protocols. *Computer Networks*, *34*(2), 317–337. doi:10.1016/S1389-1286(00)00089-X

Holland, J. H. (1992). Genetic algorithms. *Scientific American*, *267*(1), 66–73. doi:10.1038cie ntificamerican0792-66

Hope, A., Kelleher, C., Holmes, L., & Hennessy, T. (1999). Health and safety practices among farmers and other workers: A needs assessment. *Occupational Medicine (Philadelphia, Pa.)*, *49*, 231–235. PMID:10474914

Hsu, C.-J., Huang, C.-Y., & Chang, J.-R. (2011). Enhancing software reliability modeling and prediction through the introduction of time-variable fault reduction factor. *Applied Mathematical Modelling*, *35*(1), 506–521. doi:10.1016/j.apm.2010.07.017

Huang, L., Yuan, F., & Xu, Q., (2009). Lifetime Reliability-Aware Task Allocation and Scheduling for MPSoC Platform. *Design, Automation and Test in Europe Conference & Exhibition*, 51-56.

Huang, C.-Y., & Lyu, M. R. (2005). Optimal release time for software systems considering cost, testing-effort, and test efficiency. *IEEE Transactions on Reliability, 54*(4), 583–591. doi:10.1109/TR.2005.859230

Hughes, A. (1993). *Electric Motors and Drives*. Oxford, UK: Newness.

Huiyun, X., Zengzhen, W., Lorann, S., Thomas, J. K., Xuzhen, H., & Xianghua, F. (2000). Agricultural work-related injuries among farmers in Hubei, People's Republic of China. *American Journal of Public Health, 90*(8), 1269–1276. doi:10.2105/AJPH.90.8.1269 PMID:10937008

Huiyun, X., Zengzhen, W., Lorann, S., Thomas, J. K., Xuzhen, H., & Xianghua, F. (2000). Article. *American Journal of Public Health, 90*, 1269–1276.

IEEE Standard Glossary of Software Engineering Terminology. (1990). *IEEE Std 610.12-1990*. New York, NY: Standards Coordinating Committee of the Computer Society of IEEE.

Inoue, S., & Yamada, S. (2008). *Optimal software release policy with change-point*. Paper presented at the International Conference on Industrial Engineering and Engineering Management, Singapore. 10.1109/IEEM.2008.4737925

Jadhav, A. N., & Gomathi, N. (2018). WGC: hybridization of exponential grey wolf optimizer with whale optimization for data clustering. *Alexandria Engineering Journal, 57*(3), 1569-1584.

Jain, M., Manjula, T., & Gulati, T. (2012). Software reliability growth model (SRGM) with imperfect debugging, fault reduction factor and multiple change-point. *Proceedings of the International Conference on Soft Computing for Problem Solving (SocProS 2011)*.

Jayabarathi, T., Raghunathan, T., Adarsh, B., & Suganthan, P. N. (2016). Economic dispatch using hybrid grey wolf optimizer. *Energy, 111*, 630–641. doi:10.1016/j.energy.2016.05.105

Jelinski, Z., & Moranda, P. (1972). *Software reliability research. In Statistical computer performance evaluation* (pp. 465–484). Elsevier. doi:10.1016/B978-0-12-266950-7.50028-1

Jonathan, R. J., Kozak, M., & Friedman, E. G., (2003). A 0.8 Volt High Performance OTA Using Bulk-Driven MOSFETs for Low Power Mixed-Signal SOCs. *IEEE International Systems-on-Chip, SOC Conference*.

Jones, C. (1996). Software defect-removal efficiency. *Computer, 29*(4), 94–95. doi:10.1109/2.488361

Jung, W. (2005). *Op Amp Applications Handbook*. Newnes.

Kachuee, M., Kiani, M. M., Mohammadzade, H., & Shabany, M. (2015, May). Cuff-less high-accuracy calibration-free blood pressure estimation using pulse transit time. In 2015 IEEE international symposium on circuits and systems (ISCAS) (pp. 1006-1009). IEEE. doi:10.1109/ISCAS.2015.7168806

Kandasamy, N., Hayes, J. P., & Murray, B. T. (2003). Dependable Communication Synthesis for Distributed Embedded Systems. *IEEE Transactions on Computers, 52*(5), 113–125. doi:10.1109/TC.2003.1176980

Kandemir, M., Li, F., Chen, G., Chen, G., & Ozturk, O. (2005). Studying storage recomputation tradeoffs in memory-constrained embedded processing. *Proc. of the Conf. on Design, Automation and Test in Europe*, 1026-1031 10.1109/DATE.2005.285

Kanoun, K. (1989). *Software dependability growth: characterization, modeling, evaluation.* Doctorat ès-Sciences thesis, Institut National polytechnique de Toulouse, LAAS report (89-320).

Kapur, P. K., Yamada, S., Aggarwal, A. G., & Shrivastava, A. K. (2013). Optimal price and release time of a software under warranty. *International Journal of Reliability Quality and Safety Engineering*, *20*(03), 1340004. doi:10.1142/S0218539313400044

Kapur, P., Aggarwal, A. G., Kapoor, K., & Kaur, G. (2009). Optimal testing resource allocation for modular software considering cost, testing effort and reliability using genetic algorithm. *International Journal of Reliability Quality and Safety Engineering*, *16*(06), 495–508. doi:10.1142/S0218539309003538

Kapur, P., & Garg, R. (1992). A software reliability growth model for an error-removal phenomenon. *Software Engineering Journal*, *7*(4), 291–294. doi:10.1049ej.1992.0030

Kapur, P., Pham, H., Aggarwal, A. G., & Kaur, G. (2012). Two dimensional multi-release software reliability modeling and optimal release planning. *IEEE Transactions on Reliability*, *61*(3), 758–768. doi:10.1109/TR.2012.2207531

Kapur, P., Pham, H., Anand, S., & Yadav, K. (2011). A unified approach for developing software reliability growth models in the presence of imperfect debugging and error generation. *IEEE Transactions on Reliability*, *60*(1), 331–340. doi:10.1109/TR.2010.2103590

Kapur, P., & Younes, S. (1996). Modelling an imperfect debugging phenomenon in software reliability. *Microelectronics and Reliability*, *36*(5), 645–650. doi:10.1016/0026-2714(95)00157-3

Karaboga, D., & Basturk, B. (2007). A powerful and efficient algorithm for numerical function optimization: Artificial bee colony (ABC) algorithm. *Journal of Global Optimization*, *39*(3), 459–471. doi:10.100710898-007-9149-x

Karima, G. K., Hassen, N., Ettaghzouti, T., & Besbes, K. (2018). A low voltage and low power OTA using Bulk-Driven Technique and its application in Gm-c Filter. *Multi-Conference on Systems, Signals & Devices*, 429-434.

Karthikeyan, C., Veeraragavathatham, D., Karpagam, D., & Ayisha, F. S. (2009). Traditional tools in agricultural practices. *Indian Journal of Traditional Knowledge*, *8*(2), 212–217.

Khan, R. H. (2014). *Performance and Performability Modeling Framework Considering Management of Service Components Deployment* (PhD Thesis). Norwegian University of Science and Technology, Trondheim, Norway.

Khan, R. H., Machida, F., Heegaard, P., & Trivedi, K. S. (2012a). From UML to SRN: A performability modeling framework considering service components deployment. In *Proceeding of the International Conference on Network and Services* (pp. 118-127). St. Marteen: IARIA.

Khan, R. H., & Heegaard, P. (2011). A Performance modeling framework incorporating cost efficient deployment of multiple collaborating components. In *Proceedings of the International Conference Software Engineering and Computer Systems* (pp. 31-45). Pahang, Malaysia: Springer. 10.1007/978-3-642-22170-5_3

Khan, R. H., & Heegaard, P. (2014). Software Performance evaluation utilizing UML Specification and SRN model and their formal representation. *Journal of Software*, *10*(5), 499–523. doi:10.17706/jsw.10.5.499-523

Khan, R. H., Machida, F., & Heegaard, P., & Trivedi, K. S. (2012b). From UML to SRN: A performability modeling framework considering service components deployment. *International Journal on Advances in Networks and Services*, *5*(3&4), 346–366.

Kilani, D., Alhawari, M., Mohammad, B., Saleh, H., & Ismail, M. (2016). An Efficient Switched-Capacitor DC-DC Buck Converter for Self-Powered Wearable Electronics. *IEEE Transactions on Circuits and Systems. I, Regular Papers*, 1–10.

Kim, T., Lee, K., & Baik, J. (2015). An effective approach to estimating the parameters of software reliability growth models using a real-valued genetic algorithm. *Journal of Systems and Software*, *102*, 134–144. doi:10.1016/j.jss.2015.01.001

Kimura, M., Toyota, T., & Yamada, S. (1999). Economic analysis of software release problems with warranty cost and reliability requirement. *Reliability Engineering & System Safety*, *66*(1), 49–55. doi:10.1016/S0951-8320(99)00020-4

Klutke, G., Kiessler, P. C., & Wortman, M. A. (2015). A critical look at the bathtub curve. *IEEE Transactions on Reliability*, *52*(1), 125–129. doi:10.1109/TR.2002.804492

Knapp, L. W. (1966). Occupational and Rural Accidents. *Archives of Environmental Health*, *13*(4), 501–506. doi:10.1080/00039896.1966.10664604 PMID:5922020

Knapp, L. W. Jr. (1965). Agricultural Injuries Prevention. *Journal of Occupational Medicine.*, *7*(11), 553–745. doi:10.1097/00043764-196511000-00001 PMID:5831719

Koc, H., Shaik, S. S., & Madupu, P. P. (2019). Reliability Modeling and Analysis for Cyber Physical Systems. *Proceedings of IEEE 9th Annual Computing and Communication Workshop and Conference*, 448-451. 10.1109/CCWC.2019.8666606

Koc, H., Kandemir, M., & Ercanli, E. (2010). Exploiting large on-chip memory space through data recomputation. *IEEE International SOC Conference*, 513-518. 10.1109/SOCC.2010.5784683

Koc, H., Kandemir, M., Ercanli, E., & Ozturk, O. (2007). Reducing Off-Chip Memory Access Costs using Data Recomputation in Embedded Chip Multi-processors. *Proceedings of Design and Automation Conference*, 224-229.

Koc, H., Tosun, S., Ozturk, O., & Kandemir, M. (2006). Reducing memory requirements through task recomputation in embedded systems. *Proceedings of IEEE ComputerSociety Annual Symposium on VLSI: Emerging VLSI Technologies and Architecture.* 10.1109/ISVLSI.2006.77

Kogure, M., & Akao, Y. (1983). Quality function deployment and CWQC in Japan. *Journal of Quality and Progress, 16*(10), 25–29.

Kopetz, H., Kantz, H., Grunsteidl, G., Puschner, P., & Reisinger, J. (1990). Tolerating Transient Faults in MARS. *20th Int. Symp. On Fault-Tolerant Computing,* 466-473.

Kraemer, F. A., Slåtten, V., & Herrmann, P. (2007). Engineering Support for UML Activities by Automated Model-Checking – An Example. In *Proceedings of the 4th International Workshop on Rapid Integration of Software Engineering Techniques,* (pp. 51-66). Academic Press.

Kremer, W. (1983). Birth-death and bug counting. *IEEE Transactions on Reliability, 32*(1), 37–47. doi:10.1109/TR.1983.5221472

Kumar, A., Singh, J. K., Mohan, D., & Varghese, M. (2008). Farm hand tools injuries: A case study from northern India. *Safety Science, 46*(1), 54–65. doi:10.1016/j.ssci.2007.03.003

Kumar, A., Varghese, M., Pradhan, C. K., Goswami, A., Ghosh, S. N., & Nag, P. K. (1986). Evaluation of working with spade in agriculture. *The Indian Journal of Medical Research, 84,* 424–429. PMID:3781600

Kumar, M., Kiran, K., & Noorbasha, F. (2017). A 0.5V bulk-driven operational trans conductance amplifier for detecting QRS complex in ECG signal. *International Journal of Pure and Applied Mathematics, 117,* 147–153.

Kumar, V., & Kumar, D. (2017). An astrophysics-inspired grey wolf algorithm for numerical optimization and its application to engineering design problems. *Advances in Engineering Software, 112,* 231–254. doi:10.1016/j.advengsoft.2017.05.008

Kumar, V., Singh, V., Dhamija, A., & Srivastav, S. (2018). Cost-Reliability-Optimal Release Time of Software with Patching Considered. *International Journal of Reliability Quality and Safety Engineering, 25*(04), 1850018. doi:10.1142/S0218539318500183

Kyriakidis, M., Majumdar, A., & Ochieng, W. (2018). The human performance railway operational index-a novel approach to assess human performance for railway operations. *Reliability Engineering & System Safety, 170,* 226–243. doi:10.1016/j.ress.2017.10.012

Kyriakidis, M., Pak, K. T., & Majumdar, A. (2015). Railway accidents caused by human error: Historic analysis of UK railways, 1945 to 2012. *Transportation Research Record: Journal of the Transportation Research Board, 2476*(1), 126–136. doi:10.3141/2476-17

Lahmiri, S., & Shmuel, A. (2019). Performance of machine learning methods applied to structural MRI and ADAS cognitive scores in diagnosing Alzheimer's disease. *Biomedical Signal Processing and Control, 52,* 414–419. doi:10.1016/j.bspc.2018.08.009

Lamport. (2017). *Specifying Systems.* Addison-Wesley.

Lee, C., Kim, H., Park, H., Kim, S., Oh, H., & Ha, S. (2010). A Task Remapping Technique for Reliable Multi-Core Embedded Systems. *Int. Conference on CODES/ISSS,* 307-316. 10.1145/1878961.1879014

Lee, S.-H., & Song, J. (2016). Bayesian-network-based system identification of spatial distribution of structural parameters. *Engineering Structures, 127*, 260–277. doi:10.1016/j.engstruct.2016.08.029

Leung, Y.-W. (1992). Optimum software release time with a given cost budget. *Journal of Systems and Software, 17*(3), 233–242. doi:10.1016/0164-1212(92)90112-W

Leveson, N. (2012). *Engineering a Safer World: Systems Thinking Applied to Safety.* Cambridge, MA: MIT Press. doi:10.7551/mitpress/8179.001.0001

Lewi, E. (1995). *Introduction to Reliability Engineering.* Wiley.

Liao, W. H., Kao, Y., & Li, Y. S. (2011). A sensor deployment approach using glowworm swarm optimization algorithm in wireless sensor networks. *Expert Systems with Applications, 38*(10), 12180–12188. doi:10.1016/j.eswa.2011.03.053

Li, Q., Chen, H., Huang, H., Zhao, X., Cai, Z., Tong, C., & Tian, X. (2017). An enhanced grey wolf optimization-based feature selection wrapped kernel extreme learning machine for medical diagnosis. *Computational and Mathematical Methods in Medicine.* PMID:28246543

Li, Q., & Pham, H. (2017a). NHPP software reliability model considering the uncertainty of operating environments with imperfect debugging and testing coverage. *Applied Mathematical Modelling, 51*, 68–85. doi:10.1016/j.apm.2017.06.034

Li, Q., & Pham, H. (2017b). A testing-coverage software reliability model considering fault removal efficiency and error generation. *PLoS One, 12*(7). doi:10.1371/journal.pone.0181524 PMID:28750091

Liu, H.-W., Yang, X.-Z., Qu, F., & Shu, Y.-J. (2005). *A general NHPP software reliability growth model with fault removal efficiency.* Academic Press.

Li, X., Xie, M., & Ng, S. H. (2010). Sensitivity analysis of release time of software reliability models incorporating testing effort with multiple change-points. *Applied Mathematical Modelling, 34*(11), 3560–3570. doi:10.1016/j.apm.2010.03.006

Lombardi, M., & Fargnoli, M. (2018). Article. *International Journal of Safety and Security Engineering, 8*(2), 342–353.

Lombardi, M., & Fargnoli, M. (2018). Prioritization of hazards by means of a QFD- based procedure. *International Journal of Safety and Security Engineering., 8*(2), 342–353. doi:10.2495/SAFE-V8-N2-342-353

Lu, C., Gao, L., Li, X., & Xiao, S. (2017). A hybrid multi-objective grey wolf optimizer for dynamic scheduling in a real-world welding industry. *Engineering Applications of Artificial Intelligence, 57*, 61–79. doi:10.1016/j.engappai.2016.10.013

Lu, C., Gao, L., & Yi, J. (2018). Grey wolf optimizer with cellular topological structure. *Expert Systems with Applications, 107*, 89–114. doi:10.1016/j.eswa.2018.04.012

Luo, M., & Wu, S. (2018). A mean-variance optimisation approach to collectively pricing warranty policies. *International Journal of Production Economics, 196,* 101–112. doi:10.1016/j.ijpe.2017.11.013

Lynn, N., & Suganthan, P. N. (2017). Ensemble particle swarm optimizer. *Applied Soft Computing, 55,* 533–548. doi:10.1016/j.asoc.2017.02.007

Mainardi, Cerutti, & Bianchi. (2006). *The Biomedical Engineering Handbook, Digital Biomedical Signal Acquisition and Processing.* Taylor & Francis Group.

Malaiya, Y. K., Von Mayrhauser, A., & Srimani, P. K. (1993). An examination of fault exposure ratio. *IEEE Transactions on Software Engineering, 19*(11), 1087–1094. doi:10.1109/32.256855

Manimaran, G., & Murthy, C. S. R. (1998). A fault-tolerant dynamic scheduling algorithm for multiprocessor real-time systems and its analysis. *IEEE Transactions on Parallel and Distributed Systems, 9*(11), 1137–1152. doi:10.1109/71.735960

Mao, C., Xiao, L., Yu, X., & Chen, J. (2015). Adapting ant colony optimization to generate test data for software structural testing. *Swarm and Evolutionary Computation, 20,* 23–36. doi:10.1016/j.swevo.2014.10.003

Maritan, D., & Panizzolo, R. (2009). Identifying business priorities through quality function deployment. *Marketing Intelligence & Planning, 27*(5), 714–728.

Marseguerra, M., Zio, E., & Librizzi, M. (2006). Quantitative developments in the cognitive reliability and error analysis method (CREAM) for the assessment of human performance. *Annals of Nuclear Energy, 33*(10), 894–910. doi:10.1016/j.anucene.2006.05.003

Martz, H. F., & Waller, R. A. (1982). *Bayesian Reliability Analysis.* New York: John Wiley & Sons, Inc.

Matej, R. M., Stopjakova, V., & Arbet, D. (2017). Analysis of bulk-driven technique for low-voltage IC design in 130 nm CMOS technology. International *Conference on Emerging eLearning Technologies and Applications (ICETA).*

Matej, R. (2017). Design techniques for low-voltage analog integrated circuits. *Journal of Electrical Engineering, 68*(4), 245–255. doi:10.1515/jee-2017-0036

Mathews, O., Koc, H., & Akcaman, M. (2015). Improving Reliability through Fault Propagation Scope in Embedded Systems. *Proceedings of the Fifth International Conference on Digital Information Processing and Communications (ICDIPC2015),* 300-305. 10.1109/ICDIPC.2015.7323045

McLeod, R. W., Walker, G. H., & Moray, N. (2005). Analysing and modelling train driver performance. *Applied Ergonomics, 36*(6), 671–680. doi:10.1016/j.apergo.2005.05.006 PMID:16095554

Mech, L. D. (1999). Alpha status, dominance, and division of labor in wolf packs. *Canadian Journal of Zoology, 77*(8), 1196–1203. doi:10.1139/z99-099

Medjahed, S. A., Saadi, T. A., Benyettou, A., & Ouali, M. (2016). Gray wolf optimizer for hyperspectral band selection. *Applied Soft Computing, 40*, 178–186. doi:10.1016/j.asoc.2015.09.045

Michiel, S. M., van Roermund, A. H. M., & Casier, H. (2008). Analog Circuit Design: High-speed Clock and Data Recovery, High-performance Amplifiers, Power management. Springer.

Mirjalili, S. (2015). The ant lion optimizer. *Advances in Engineering Software, 83*, 80–98. doi:10.1016/j.advengsoft.2015.01.010

Mirjalili, S., & Lewis, A. (2016). The whale optimization algorithm. *Advances in Engineering Software, 95*, 51–67. doi:10.1016/j.advengsoft.2016.01.008

Mirjalili, S., Mirjalili, S. M., & Lewis, A. (2014). Grey wolf optimizer. *Advances in Engineering Software, 69*, 46–61. doi:10.1016/j.advengsoft.2013.12.007

Mirjalili, S., Saremi, S., Mirjalili, S. M., & Coelho, L. D. S. (2016). Multi-objective grey wolf optimizer: A novel algorithm for multi-criterion optimization. *Expert Systems with Applications, 47*, 106–119. doi:10.1016/j.eswa.2015.10.039

Mishra, D., & Satapathy, S. (2018). Drudgery Reduction of Farm Women of Odisha by Improved Serrated Sickle. *International Journal of Mechanical Engineering and Technology, 9*(2), 53–61.

Mittal, V. K., Bhatia, B. S., & Ahuja, S. S. (1996). *A Study of the Magnitude, Causes, and Profile of Victims of Injuries with Selected Farm Machines in Punjab. Final Report of ICAR adhoc Research Project*. Ludhiana, India: Department of Farm Machinery and Power Engineering, Punjab Agricultural University.

Mohamed, K. (2013). Wireless Body Area Network: From Electronic Health Security Perspective. *International Journal of Reliable and Quality E-Healthcare, 2*(4), 38–47. doi:10.4018/ijrqeh.2013100104

Mohan, D. (2000). Equipment-related injuries in agriculture: An international perspective. *Injury Control and Safety Promotion, 7*(3), 1–12.

Mohan, D., & Patel, R. (1992). Design of safer agricultural equipment: Application of ergonomics and epidemiology. *International Journal of Industrial Ergonomics, 10*(4), 301–309. doi:10.1016/0169-8141(92)90097-J

Mohanty, R., Ravi, V., & Patra, M. R. (2010). The application of intelligent and soft-computing techniques to software engineering problems: A review. *International Journal of Information and Decision Sciences, 2*(3), 233–272. doi:10.1504/IJIDS.2010.033450

Moore, H. (2007). You Can't Be a Meat-Eating Environmentalist. *American Chronicle*. Retrieved from http://www.americanchronicle.com/articles/view/24825

Mu, L., Xiao, B. P., Xue, W. K., & Yuan, Z. (2015) The prediction of human error probability based on Bayesian networks in the process of task. *2015 IEEE International Conference on Industrial Engineering and Engineering Management (IEEM)*. 10.1109/IEEM.2015.7385625

Murphy, D. J. (1992). *Safety and Health for Production Agriculture.* St. Joseph, MI: American Society of Agricultural Engineers.

Musa, J. D. (1975). A theory of software reliability and its application. *IEEE Transactions on Software Engineering, SE-1*(3), 312–327. doi:10.1109/TSE.1975.6312856

Musa, J. D. (1980). The measurement and management of software reliability. *Proceedings of the IEEE, 68*(9), 1131–1143. doi:10.1109/PROC.1980.11812

Musa, J. D. (1991). Rationale for fault exposure ratio K. *Software Engineering Notes, 16*(3), 79. doi:10.1145/127099.127121

Musa, J. D. (2004). *Software reliability engineering: more reliable software, faster and cheaper.* Tata McGraw-Hill Education.

Musa, J. D., Iannino, A., & Okumoto, K. (1987). *Software reliability: Measurement, prediction, application. 1987.* McGraw-Hill.

Nag, P. K., & Nag, A. (2004). Drudgery, accidents and injuries in Indian agriculture. *Industrial Health, 42*(4), 149–162. doi:10.2486/indhealth.42.149 PMID:15128164

Naixin, L., & Malaiya, Y. K. (1996). Fault exposure ratio estimation and applications. *Proceedings of ISSRE'96: 7th International Symposium on Software Reliability Engineering.*

Namazi, A., Safari, S., & Mohammadi, S. (2019). CMV: Clustered Majority Voting Reliability-Aware Task Scheduling for Multicore Real-Time Systems. *IEEE Transactions on Reliability, 68*(1), 187–200. doi:10.1109/TR.2018.2869786

Nasar, M., Johri, P., & Chanda, U. (2013). A differential evolution approach for software testing effort allocation. *Journal of Industrial and Intelligent Information, 1*(2).

Nikhil, Ranitesh, & Vikram. (2010). Bulk driven OTA in 0.18 micron with high linearity. *International Conference on Computer Science and Information Technology.*

Nilsson, K., Pinzke, S., & Lundqvist, P. (2010). Occupational Injuries to Senior Farmers in Sweden. *Journal of Agricultural Safety and Health, 16*(1), 19–29. doi:10.13031/2013.29246 PMID:20222268

Nye, W., Riley, D. C., & Sangiovanni-Vincentelli, A. L. (1988). DELIGHT.SPICE: An optimization-based system for the design of integrated circuits. *IEEE Transactions on Computer-Aided Design of Integrated Circuits and Systems, 7*(4).

O'Connor, P., & Kleyner, A. (2012). *Practical Reliability Engineering (5th ed.).* Wiley & Sons.

OMG. (2008). *UML Profile for Modeling Quality of Service and Fault Tolerance Characteristics and Mechanism.* Version 1.1. In Object management group. Retrieved January 2019, from https://pdfs.semanticscholar.org/e4bc/ffb49b6bd96d8df35173704292866e0623eb.pdf

OMG. (2017). *OMG Unified Modeling Language™ (OMG UML) Superstructure.* Version 2.5. In Object management group. Retrieved January 2019 from http://www.omg.org/spec/UML/2.2/ Superstructure/PDF/

OMG. (2019). *UML Profile for MARTE: Modeling and analysis of real-time embedded systems.* Version 1.2. In Object management group. Retrieved January 2019, from http://www.omg.org/ omgmarte/ Documents/ Specifications/08-06-09.pdf

Pachauri, B., Dhar, J., & Kumar, A. (2015). Incorporating inflection S-shaped fault reduction factor to enhance software reliability growth. *Applied Mathematical Modelling, 39*(5-6), 1463–1469. doi:10.1016/j.apm.2014.08.006

Pai, P.-F. (2006). System reliability forecasting by support vector machines with genetic algorithms. *Mathematical and Computer Modelling, 43*(3-4), 262–274. doi:10.1016/j.mcm.2005.02.008

Panwar, L. K., Reddy, S., Verma, A., Panigrahi, B. K., & Kumar, R. (2018). Binary grey wolf optimizer for large scale unit commitment problem. *Swarm and Evolutionary Computation, 38,* 251–266. doi:10.1016/j.swevo.2017.08.002

Pasquini, A., Rizzo, A., & Save, L. (2004). A methodology for the analysis of SPAD. *Safety Science, 42*(5), 437–455. doi:10.1016/j.ssci.2003.09.010

Patel, S. K. (2010). Agricultural injuries in Etawah district of Uttar Pradesh in India. *Safety Science, 48,* 222–229. https://doi.org/10.1016/j.ssci.2009.08.003

Patel, S. K., Varma, M. R., & Kumar, A. (2010). Agricultural injuries in Etawah district of Uttar Pradesh in India. *Safety Science, 48*(2), 222–229. doi:10.1016/j.ssci.2009.08.003

Pawitra, T. A., & Tan, K. C. (2003). Tourist satisfaction in Singapore- a perspective from Indonesian tourists. *Managing Service Quality, 13*(5), 339–411.

Pham, H. (1996). A software cost model with imperfect debugging, random life cycle and penalty cost. *International Journal of Systems Science, 27*(5), 455–463. doi:10.1080/00207729608929237

Pham, H. (2014). A new software reliability model with Vtub-shaped fault-detection rate and the uncertainty of operating environments. *Optimization, 63*(10), 1481–1490. doi:10.1080/023 31934.2013.854787

Pham, H., Nordmann, L., & Zhang, Z. (1999). A general imperfect-software-debugging model with S-shaped fault-detection rate. *IEEE Transactions on Reliability, 48*(2), 169–175. doi:10.1109/24.784276

Pham, H., & Zhang, X. (1999). A software cost model with warranty and risk costs. *IEEE Transactions on Computers, 48*(1), 71–75. doi:10.1109/12.743412

Plioutsias, A., & Karanikas, N. (2015). Using STPA in the evaluation of fighter pilots training programs. *Procedia Engineering, 128,* 25–34. doi:10.1016/j.proeng.2015.11.501

Pop, P., Izosimov, V., Eles, P., & Peng, Z. (2009). Design Optimization of Time-and Cost Constrained Fault-Tolerant Embedded Systems With Checkpointing and Replication. *IEEE Trans. on VLSI Systems*, 389-401.

Pop, P., Poulsen, K. H., Izosimov, V., & Eles, P. (2007). Scheduling and Voltage Scaling for Energy/ Reliability Trade-offs in Fault-Tolerant Time-Triggered Embedded Systems. *5th International Conference on CODES+ISSS*, 233-238. 10.1145/1289816.1289873

Prasanna Kumar, G. V., & Dewangan, K. N. (2009). Agricultural accidents in north eastern region of India. *Safety Science*, *47*, 199–205. https://doi.org/10.1016/j.ssci.2008.03.007

Prasanna Kumar, G. V., & Dewangan, K. N. (2009). Article. *Safety Science*, *47*, 199–205. doi:10.1016/j.ssci.2008.03.007

Raj, N., Sharma, R. K., Jasuja, A., & Garg, R. (2010). A Low Power OTA for Biomedical Applications. *Journal of Selected Areas in Bioengineering*.

Raj, S. R., Bhaskar, D. R., Singh, A. K., & Singh, V. K. (2013). Current Feedback Operational Amplifiers and Their Applications. In Analog Circuits and Signal Processing. Springer Science & Business Media.

Rajora, Zou, Guang Yang, Wen Fan, Yi Chen, Chieh Wu, … Liang. (2016). A split-optimization approach for obtaining multiple solutions in single-objective process parameter optimizatio. *SpringerPlus*.

Raj, T., & Attri, R. (2011). Identification and modeling of barriers in the implementation of TQM. *International Journal of Productivity and Quality Management.*, *28*(2), 153–179. doi:10.1504/ IJPQM.2011.041844

Ramirez-Angulo, J., Lopez-Martin, A.J., Carvajal, R.G., & Galan, J.A. (2002). A free but efficient class AB two-stage operational amplifier. *ISCAS*.

Rausand, M. (2011). *Risk Assessment: Theory, Methods, and Applications*. John Wiley & Sons. doi:10.1002/9781118281116

Rautiainen, R. H., & Reynolds, S. J. (2002). Mortality and morbidity in agriculture in the United States. *Journal of Agricultural Safety and Health*, *8*(3), 259–276. doi:10.13031/2013.9054 PMID:12363178

Ravelomanantsoa, A. (2015). *Deterministic approach of compressed acquisition and signals reconstruction from distributed intelligent sensors* (Thesis). University of Lorraine.

Ravi, V., & Shankar, R. (2005). Analysis of interactions among the barriers of reverse logistics. *Technological Forecasting and Social Change*, *72*(8), 1011–1029. doi:10.1016/j. techfore.2004.07.002

Rekabi, M. M. (2018). *Bayesian Safety Analysis of Railway Systems with Driver Errors* (Master thesis). Norwegian University of Science and Technology.

Yamada, S., Tokuno, K., & Osaki, S. (1992). Imperfect debugging models with fault introduction rate for software reliability assessment. *International Journal of Systems Science, 23*(12), 2241–2252. doi:10.1080/00207729208949452

Yang, G. (2007). *Life Cycle Reliability Engineering*. Wiley. doi:10.1002/9780470117880

Yang, X. S. (2010). Firefly algorithm, Levy flights and global optimization. In *Research and development in intelligent systems XXVI* (pp. 209–218). London: Springer. doi:10.1007/978-1-84882-983-1_15

Yang, X. S. (2011). Algorithm for multi-objective optimization. *International Journal of Bio-inspired Computation, 3*(5), 267–274. doi:10.1504/IJBIC.2011.042259

Yavari, M., & Shoaei, O. (2005). A Novel Fully-Differential Class AB Folded Cascode OTA for Switched-Capacitor Applications. *ICECS*.

Zhang, S., & Zhou, Y. (2017). Template matching using grey wolf optimizer with lateral inhibition. *Optik (Stuttgart), 130*, 1229–1243. doi:10.1016/j.ijleo.2016.11.173

Zhang, X., Teng, X., & Pham, H. (2003). Considering fault removal efficiency in software reliability assessment. *IEEE Transactions on Systems, Man, and Cybernetics. Part A, Systems and Humans, 33*(1), 114–120. doi:10.1109/TSMCA.2003.812597

Zhang, Y., Dick, R., & Chakrabarty, K. (2004). Energy-aware deterministic fault tolerance in distributed real-time embedded systems. *Proceedings of the 41st annual Conference on Design Automation*, 550-555. 10.1145/996566.996719

Zhou, C., & Roseman, J. M. (1994). Agricultural injuries among a population based sample of farm operators in Alabama. *American Journal of Industrial Medicine, 25*(3), 385–402. doi:10.1002/ajim.4700250307 PMID:8160657

Zhou, C., & Roseman, J. M. (1994). Agricultural injuries among a population-based sample of farm operators in Alabama. *American Journal of Industrial Medicine, 25*(3), 385–402. doi: 10.1002/ajim.4700250307

Zhu, D., Melhem, R., & Mosse, D. (2004). The Effects of Energy Management on Reliability in Real-Time Embedded Systems. *International Conference on ICCAD*, 35-40.

Ziegler, J. F., Muhlfeld, H. P., Montrose, C. J., Curtis, H. W., O'Gorman, T. J., & Ross, J. M. (1996). Accelerated testing for cosmic soft-error rate. *IBM Journal of Research and Development, 40*(1), 51–72. doi:10.1147/rd.401.0051

About the Contributors

Mohamed Arezki Mellal has a PhD in Mechatronics. He is an Associate Professor (with Accreditation to Supervise Research) at the Department of Mechanical Engineering, Faculty of Technology, M'Hamed Bougara University, Algeria. He has published in several journals such as: Reliability Engineering & System Safety, Energy, International Journal of Advanced Manufacturing Technology, ISA Transactions, Journal of Intelligent Manufacturing and conferences. He was also a committee member of over sixty international conferences. He serves as a regular reviewer for several SCI journals. His research interests include mechatronics and soft computing methods for engineering problems. He undertaken several research visits and undergo a visiting scholar at the University of Maryland (USA).

* * *

Anu G. Aggarwal is working as Professor in the Department of Operational Research, University of Delhi. She obtained her Ph.D., M.Phil and M.Sc. degrees in Operational Research from the University of Delhi in year 2007, 1999 and 1996, respectively. She has published several papers in the area of Marketing Management and Theory of Reliability. Her research interests include modeling and optimization in Consumer Buying Behavior, Innovation-Diffusion modeling and Soft Computing Techniques. She has reviewed many research papers for reputed journals including IEEE Transactions on Engineering Management, International journal of System Science, Int. J. of Production Research, Int. J. of Operational Research, etc.

Nabanita Banerjee is an assistant Professor in Electronics and Instrumentation department at Techno Main, Salt-Lake. She did her M.Tech degree in Control and Instrumentation in the year 2008.

Houda Daoud was born in Sfax, Tunisia in 1980. She received the Electrical Engineering Diploma then the Master degree in electronics from the National School of Engineering of Sfax "ENIS", respectively, in 2004 and 2005. She joints the Electronic and Information Technology Laboratory of Sfax "LETI" since 2004 and she has been Assistant professor at the National School of Electronic and telecommunication "ENET'COM" from 2012. Her current research interests are on analogue CMOS integrated circuits design for biomedical applications. She is author and co-author of several journal papers, conference papers and book chapters.

Razib Hayat Khan currently works as an Assistant Professor at Department of Computer Science, American International University-Bangladesh (AIUB), Dhaka, Bangladesh. Prior joining AIUB, he worked as a Software Developer at Intelliview and PatientSky, Oslo, Norway. He finished his PhD from Norwegian University of Science and Technology (NTNU) at Trondheim, Norway in Software Engineering and completed his post-graduation form The Royal Institute of Technology (KTH) at Stockholm, Sweden in Information and Communication Systems Security. He worked as a visiting researcher at Duke University, North Carolina, USA, consultant at Ericsson, Lulea, Sweden and project member at University of Twente, Twente, Netherlands. He published more than 30 publications in International peer reviewed journals and conferences. He has given numerous talks in international conferences. He served as a Technical Program Committee member and reviewer in several international conferences and journals.

Hakduran Koc is currently chairman and associate professor of Computer Engineering at University of Houston - Clear Lake. After receiving his B.S. degree in Electronics Engineering from Ankara University, he worked in the industry for two years. Then, he joined Syracuse University where he received his M.S. and Ph.D. degrees in Computer Engineering. During his graduate study, Dr. Koc was at The Pennsylvania State University as visiting scholar. His research and teaching are in the areas of digital design, embedded systems, and computer architecture. He is the recipient of several teaching and leadership awards including UHCL Piper Award Nominee, IEEE Outstanding Student Branch Counselor Award and IEEE MGA Outstanding Small Section Award. He is currently a member of IEEE and serves as ABET Program Evaluator.

Dalila Laouej was born in Sfax, Tunisia in 1988. She received the National Diploma in Applied License in Technologies of Electronics and Communications, then the Master degree in electronics from the national school of electronics and communications of Sfax (ENET'com), respectively, in 2011 and 2014. She joints the electronic and information technology laboratory of Sfax "LETI" since 2015. Her current research interests are on analog CMOS integrated circuit design system in biomedical dispositive.

Yiliu Liu is an associate professor in the RAMS (reliability, availability, maintainability and safety) group, at Department of Mechanical and Industrial Engineering, Norwegian University of Science and Technology. His main research interests include system reliability and safety analysis, prognosis and system health management.

Mourad Loulou was born in Sfax, Tunisia in 1968. He was graduated in Electrical Engineering from the University of Sfax in 1993 and received the Award of President of the Tunisian Republic. In 1998 he obtained the Ph.D degree in Electronics from the University of Bordeaux, France. Since October 2009 he became a Full Professor and he currently leads the Analogue and RF IC Design Group at the National School of Engineers in Sfax. He is also a consulting expert in IC design. He is IEEE senior member. He is the organizer of the IEEE Tunisia Section and Section Chair from 2009 to 2012. He is also the organizer and he chaired the Tunisia section Circuits And Systems Society chapter from 2009 to 2016. He served as general co-chair, TPC co-chair and committee member of several IEEE conferences. Professor Mourad Loulou, is advisor of several Ph. D. thesis in the field of the design and the design automation of Analogue, mixed mode and RF circuits. He is author and co-author of several journal papers, conference papers, book chapters and books.

Jihene Mallek was born in Sfax, Tunisia, on September 1980. She received the Electrical Engineering Diploma then the Master degree in electronics from the National School of Engineering of Sfax, respectively, in 2006 and 2007. She joints the Electronic and Information Technology Laboratory of Sfax since 2006 and he has been a PhD student at the National School of Engineering of Sfax from 2006. She received her Ph.D. degree in 2017 in electronics from the same institution. His current research interests are on continuous time sigma delta ADC.

Oommen Mathews is currently working as Software Engineer Specialist at one of the largest independent oil & gas producers in United States. He finished his bachelors in Electronics & Communication Engineering from Karunya University, Coimbatore, India. He also completed his graduate studies in Computer Engineering from University of Houston-Clear Lake.

Sumitra Mukhopadhyay received the Master Degree and Ph.D degree, both from Jadavpur University in electronics and communication engineering in 2004 and 2009 respectively. She is currently an Assistant Professor with the University of Calcutta. Her main research interests include embedded system, field-programmable gate array based prototype design, optimization, soft computing, evolutionary computation.

Neha Neha is a Ph.D. scholar in the Department of Operational Research, University of Delhi, India. She has done her M.Phil. degree in Operational Research from the University of Delhi and has done her M.Sc in Applied Operational Research, Department of Operational Research, University of Delhi. Her research area is Software Reliability and Operational Research.

Vibha Verma is a Ph.D. scholar in the Department of Operational Research, University of Delhi, India. She has done her M.Phil. degree in Operational Research from the University of Delhi and has done her M.Sc in Applied Operational Research, Department of Operational Research, University of Delhi. Her research area is Software Reliability and Operational Research.

Seongwoo Woo has a BS and MS in Mechanical Engineering, and he has obtained PhD in Mechanical Engineering from Texas A&M. He major in energy system such as HVAC and its heat transfer, optimal design and control of refrigerator, reliability design of thermal components, and failure Analysis of thermal components in marketplace using the Non-destructive such as SEM & XRAY. In 1992.03–1997 he worked in Agency for Defense Development, Chinhae, South Korea, where he has researcher in charge of Development of Naval weapon System. Now he is working as a Senior Reliability Engineer in Side-by-Side Refrigerator Division, Digital Appliance, SAMSUNG Electronics, and focus on enhancing the life of refrigerator as using the accelerating life testing. He also has experience about Side-by-Side Refrigerator Design for Best Buy, Lowe's, Cabinet-depth Refrigerator Design for General Electrics.

Index

Ensure Quality Research is Introduced to the Academic Community

Become an IGI Global Reviewer for Authored Book Projects

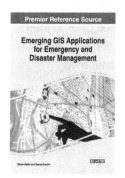
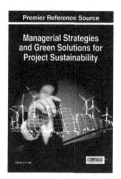

The overall success of an authored book project is dependent on quality and timely reviews.

In this competitive age of scholarly publishing, constructive and timely feedback significantly expedites the turnaround time of manuscripts from submission to acceptance, allowing the publication and discovery of forward-thinking research at a much more expeditious rate. Several IGI Global authored book projects are currently seeking highly-qualified experts in the field to fill vacancies on their respective editorial review boards:

Applications and Inquiries may be sent to:
development@igi-global.com

Applicants must have a doctorate (or an equivalent degree) as well as publishing and reviewing experience. Reviewers are asked to complete the open-ended evaluation questions with as much detail as possible in a timely, collegial, and constructive manner. All reviewers' tenures run for one-year terms on the editorial review boards and are expected to complete at least three reviews per term. Upon successful completion of this term, reviewers can be considered for an additional term.

If you have a colleague that may be interested in this opportunity, we encourage you to share this information with them.